The Scholarship Series in Biology

General Editor: W. H. Dowdeswell

Enzymes and Equilibria

THE SCHOLARSHIP SERIES IN BIOLOGY

Human Blood Groups and Inheritance
S. Lawler and L. J. Lawler

Woodland Ecology
E. G. Neal

New Concepts in Flowering Plant Taxonomy
J. Heslop Harrison

The Mechanism of Evolution
W. H. Dowdeswell

Heathland Ecology
C. P. Friedlander

The Organization of the Central Nervous System
C. V. Brewer

Plant Metabolism
G. A. Strafford

Comparative Physiology of Vertebrate Respiration
G. M. Hughes

Animal Body Fluids and their Regulation
A. P. M. Lockwood

Life on the Sea-shore
A. J. Southward

Chalkland Ecology
John Sankey

Ecology of Parasites
N. A. Croll

Physiological Aspects of Water and Plant Life
W. M. M. Baron

Biological Principles in Fermentation
J. G. Carr

Studies in Invertebrate Behaviour
S. M. Evans

Ecology of Refuse Tips
Arnold Darlington

Plant Growth
Michael Black and Jack Edelman

Animal Taxonomy
T. H. Savory

Ecology of Fresh Water
Alison Leadley Brown

Ecology of Estuaries
Donald S. McLusky

Variation and Adaptation in Plant Species
David A. Jones and Dennis A. Wilkins

Nature Reserves and Wildlife
Eric Duffey

Enzymes and Equilibria
J. C. Marsden and C. F. Stoneman

Enzymes and Equilibria

J. C. Marsden
Reader in Cell Physiology, Sir John Cass College of Science and Technology, City of London Polytechnic

and

C. F. Stoneman
Lecturer in Education, University of York

Heinemann Educational Books

Heinemann Educational Books Ltd

LONDON EDINBURGH MELBOURNE AUCKLAND TORONTO
HONG KONG SINGAPORE KUALA LUMPUR
IBADAN NAIROBI JOHANNESBURG
LUSAKA NEW DELHI

ISBN 0 435 61840 7
© J. C. Marsden and C. F. Stoneman 1974
First published 1974

Published by
Heinemann Educational Books Ltd
48 Charles Street, London W1X 8AH

Printed in Great Britain by
The Whitefriars Press Ltd
London and Tonbridge

Preface

A little book cannot possibly do justice to a subject as vast and important as the nature of enzymes. It must either be merely introductory and remain elementary or deal with a restricted number of topics. This book attempts the latter—to discuss a few important points connected with enzymes in biological systems. The selection of these aspects of enzymology is based on experience with students in the first year of biology degree courses. The book is intended for those who have already met enzymes at school level and wish to progress from elementary, and perhaps over-simplified notions, to consider more difficult parts of the subject which are still, to a large extent, conjectural and speculative. Attempts at understanding enzyme action at molecular level present a considerable intellectual challenge. The results of research sometimes confirm a model of enzyme structure or function but, often enough, ideas are upset and new models have to be invented. If the reader is stimulated by uncertainty about some features of enzymes and informed about others, this book will have served its purpose.

We are most grateful to Dr A W Robards and Mr Peter Crosby, of the University of York, for providing electron micrographs of plant and animal material, to Dr Olga Kennard, of the Crystallography Laboratory, Cambridge, for the photograph of an ATP model, to Dr D G Hoare of York University for the use of his chymotrypsin model, and to Dr D M Blow of the MRC Laboratory of Molecular Biology, Cambridge, for advice on the representation of chymotrypsin structure and his permission, with that of his associates, Dr B W Matthews, Dr P B Sigler, and Dr R Henderson, to use material first publised in *Nature*.

J.C.M.

C.F.S.

1974

Contents

	page
Preface	v
1. The Living Cell	1
2. Atoms and Molecules	17
3. Enzymes, Equilibria, and Energy	28
4. Enzyme Activity	45
5. Enzyme Models	63
6. Oxidation and Reduction	76
7. The Significance of Enzymes	86
8. Techniques of Investigation	94
Suggestions for Further Reading	112
Index	113

1.

The Living Cell

An organism contains a vast number of complex chemical reactions which are sustained during its life and that of its progeny. There are many different kinds of living creatures, varying in form from micro-organisms, like bacteria, to the so-called higher plants and animals. By 'higher' we mean more complex and showing a greater degree of organization. The diversity of form is truly amazing. It has been estimated that there are about 1 300 000 distinct species of living organisms on this planet at the present time; many more thousands have become extinct at some time in the past.

It might seem reasonable to suppose that the chemistry going on inside organisms differs as widely as the external form or appearance, but this is not so. At first sight the common bacterium *Escherichia coli,* which appears to be little more than a minute bag of jelly, 1 μm in diameter, bears little resemblance to, say, a rat 180 mm long, with a complex shape and a mass 4×10^{13} times greater. But one most remarkable result of seventy years of biochemical investigation has been the realization that two such apparently different life forms have much in common at the chemical or molecular level. Indeed, at this level we are uncertain about why they are so different! *E. coli* is an acellular organism, a single unit of protoplasm surrounded by one membrane. Inside this membrane is a fluid of great chemical and structural complexity, within which occurs a set of chemical processes which make *E. coli* a unique species. Many of its reactions are the same as those of other organisms but some are peculiar and these distinguish it from other species.

The rat, on the other hand, is made up of billions of cells. Individual cells of the rat (Figure 1.1), though bigger than a single *E. coli* bacterium, are sometimes similar in outline to them, but rat cells show more diversity of form and more internal complexity.

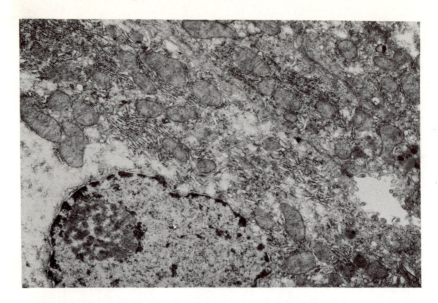

Figure 1.1. Electron micrograph of part of a cell from a salivary gland of a rat showing something of the structural complexity within a single cell (magnification x15 000)

Ultrastructure

Membranes

The minute, internal features of cells are relevant to a study of cell chemistry and it is therefore necessary to review cell ultrastructure before describing biochemical reactions which take place in them.

The cell membrane or plasmalemma is a convenient starting point, since it is common to all organisms capable of independent existence. The membrane represents a physical separation of the cell from its environment. It must allow all the materials necessary for life to pass in and all poisonous products to pass out. But, at the same time it must keep the protoplasm together, for if it became widely dispersed and diluted most chemical reactions would cease. The importance of bounding membranes in the life of a cell is very great, and how they manage to perform these functions has been the subject of much investigation. Current views on cell membranes suggest that it is composed partly of fatty material (see Figure 1.2) to stabilize it in a watery environment and partly of protein which provides specific carriers for the transport of nutrients and waste products through from one side to the other. This fragile structure, only about 10 nm thick, is responsible for importing dozens of chemical components to enable a bacterium to survive. For the more sophisticated mammal cell the figure may run to hundreds.

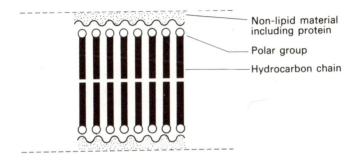

Non-lipid material including protein

Polar group

Hydrocarbon chain

Figure 1.2. Part of a cell membrane represented diagrammatically as a lipid bilayer

Some substances have to be 'pumped' in or out of the cell from a low concentration to a higher one, a process requiring energy. However, it is important to note that once a mechanism for transporting one material across the cell membrane of one species has been established then it is reasonable to expect that a similar mechanism will apply elsewhere, and although the organization of many different carriers within a single membrane may be exceedingly complex the complexity arises from a multiplication of simpler processes. Whilst these may vary from species to species, the underlying mechanism is a fundamental property of the membranes of all living organisms.

In addition to a membrane, plant cells and some acellular organisms such as *E. coli* have a cell wall (see Figure 1.3) which protects the flimsy membrane underneath. In plants it consists largely of the carbohydrate cellulose. The toughness of this material has been known and exploited by man for a long time; in acellular organisms it takes a more complex chemical form.

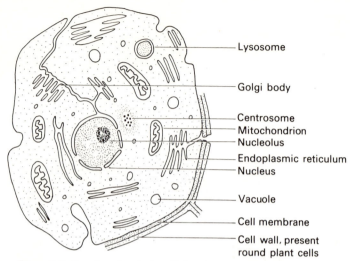

Lysosome

Golgi body

Centrosome
Mitochondrion
Nucleolus
Endoplasmic reticulum
Nucleus

Vacuole

Cell membrane

Cell wall, present
round plant cells

Figure 1.3. Drawing of a single cell showing several common features

Nuclei

The membrane is a common feature of all life forms; the nucleus is found only in some organisms—the eukaryotic species—which represent an extra degree of complexity over the simpler prokaryotes (which lack nuclei). The nucleus is the site, within the cell, of most of the genetic material—the deoxyribonucleic acid (DNA). Its function is to carry information to program a cell so that it carries out all its chemical tasks correctly. The DNA structure can be copied so that division results in two daughter cells each with the same genetic content. Though generally stable, DNA can undergo change or mutation so providing a source of variation within a species so that selection and evolution can occur.

The nucleus is separated from the cytoplasm by a membrane thinner than the cell membrane but effective in preventing the dispersal of nuclear contents which would cause difficulties at times of cell division.

Prokaryotes, which include the bacterium *E. coli,* do not have nuclei but they do possess DNA scattered through the protoplasm. Dispersal of genetic material does not seem to be a disadvantage in such simple organisms and there is very much less of it than in eukaryotic creatures. It is perhaps worth emphasizing that almost all our knowledge of the structure and function of DNA in life has been obtained from studies of prokaryotes The eukaryotic organisms do not lend themselves so well to these studies because their growth rates are slower and their nuclei are more complicated. Though the external appearances differ, the underlying biochemical processes associated with DNA, whether enclosed in a nucleus or not, are shared nevertheless by all forms of life.

Ribosomes

The cell membrane and the nucleus are distinctive cellular features or organelles. Others include the Golgi complex, endoplasmic reticulum, mitochondria, lysosomes, and ribosomes. These last are common to every organism and appear as small dots in electron micrographs. They are too small to be studied in detail under an electron microscope but too large for techniques such as X-ray diffraction. So almost all our knowledge of these organelles derives from indirect biochemical observations. The ribosome is the site of protein synthesis in the cell. Proteins are complex substances composed of many much simpler chemicals called amino acids. The structural formulae of four amino acids are given below:

$$
\begin{array}{llll}
NH_2 & NH_2 & NH_2 & NH_2 \\
| & | & | & | \\
HCH & HC{-}CH_3 & HC{-}CH_2{-}SH & HC{-}CH_2{-}CH_2{-}CH_2{-}N^+H_3 \\
| & | & | & | \\
COOH & COOH & COOH & COO^- \\
\text{glycine} & \text{alanine} & \text{cysteine} & \text{ornithine (zwitterion)}
\end{array}
$$

The following shows the way in which amino acids combine (R, R' etc. represent parts of different amino acids).

$$
\begin{array}{ccccccc}
R & O & & R'' & O \\
| & \| & H & | & \| \\
CH & C & N & CH & C \\
\diagup \diagdown & \diagup \diagdown & \diagup \diagdown & \diagup \diagdown \\
N & CH & C & N & CH \\
H & | & \| & H & | \\
& R' & O & & R'''
\end{array}
$$

There are some twenty amino acids, combined in various configurations, which make up hundreds of different kinds of protein molecules. The ribosome can be regarded as a small factory in which the amino acids are assembled, in a predetermined sequence, into proteins. Although the ribosomes of *E. coli* are smaller than those of higher organisms, the task they perform seems to be the same. The importance of proteins in living material cannot be exaggerated; enzymes, membranes, muscle, and many other essential parts of organisms are composed largely or wholly of protein.

Mitochondria

To carry out the protein-making function the ribosomes need a supply of energy. In the cells of higher organisms this comes from the mito-chondria. These have a curious striped appearance when observed under the electron microscope. The stripes or cristae are interpreted as deeply folded inner membranes (Figure 1.4). Along these membranes are located enzymes concerned with respiration. These enzymes are responsible for the oxidation of chemicals, in the cell, producing substances capable of transferring energy from oxidation to maintain vital processes such as membrane 'pumps', protein synthesis, and cell movement. Mitochondria, unlike other organelles, replicate themselves

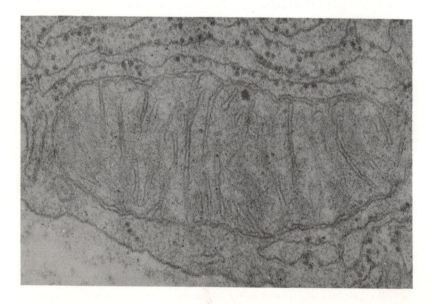

Figure 1.4. Electron micrograph of a single mitochondrion from a rat salivary gland cell showing the cristae, infoldings of the inner membrane (magnification x120 000)

independently when the cell divides and each one has its own (limited) supply of DNA. Some simple organisms are too small to contain mitochondria but they do have enzymes capable of oxidizing food material to provide useful energy. At the biochemical level the actual processes of energy supply are similar.

Chloroplasts

Plants, and some bacteria, differ from animals in that they can carry out photosynthesis, that is, use sunlight as a source of energy for life. All life on this planet depends ultimately on photosynthesis. The food consumed by non-photosynthetic organisms is produced in the first place by the photosynthetic ones. In plants, the photosynthetic process is confined to special organelles called chloroplasts which (like mito-chondria) also have an inner arrangement of membranes carrying the necessary molecules—proteins, fats, and chlorophyll—to 'fix' light for cellular chemistry or metabolic processes.

Endoplasmic reticulum

The protoplasm of prokaryotes can be regarded as a 'soup' in which the DNA, ribosomes, and many other constituents float, but the cytoplasm of the cells of higher organisms is more complex. In the two dimensions of an electron micrograph (see Figure 1.5) the endoplasmic reticulum appears as thin threads dividing the cell into many compartments, but in three dimensions these compartments should be imagined as balloon-like, bounded by the endoplasmic reticulum. It is assumed that these compartments localize particular groups of chemical reactions and help their co-ordination. Organelles such as nuclei and mitochondria can also be regarded as compartments in this sense although the membranes which surround these organelles are more complex than endoplasmic reticulum. The outer membrane (plasmalemma), that of the nucleus, and the endoplasmic reticulum are all connected in a complex manner.

   The ways in which things pass between the various compartments bounded by endoplasmic reticulum is still uncertain. Free diffusion may enable small molecules to pass through the membranes, but for larger molecules perhaps some kind of directed flow occurs, and it has been suggested that other features of ultrastructure—microtubules—play a part in this process (Figure 1.6).

Figure 1.5. Electron micrograph of part of a cow salivary gland showing endoplasmic reticulum. The thin membranes shown in section are associated with ribosomes which appear as small dots (magnification x45 000)

Microtubules are structures believed to be composed of protein sub-units and they have been seen in many kinds of cell. They seem to be easily assembled and dismantled by the cell and to control the movement of materials within it.

In some parts of cells the endoplasmic reticulum, as seen in electron micrographs, is called 'rough' due to a coating of ribosomes on it. Without this attachment to membranes, it is believed that the ribosomes of higher organisms would be incapable of protein synthesis. In other parts, the endoplasmic reticulum is described as 'smooth' where it lacks ribosomes. In cells where protein synthesis is a large part of the overall economy, for example in cells of the pancreas or salivary glands producing enzymes for digestive secretion, the proportion of rough endoplasmic reticulum is much greater than in other cells. Besides ribosomes a number of enzymes are associated with the endoplasmic reticulum and do not function properly unless they are thus associated.

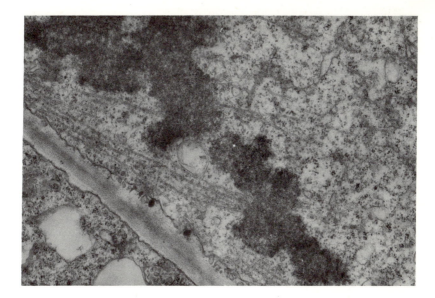

Figure 1.6. Electron micrograph of part of a plant cell *(Salix fragilis)* showing microtubules. These appear as parallel lines running diagonally (magnification x30 000)

Lysosomes

The Golgi body is a complex convergence of the endoplasmic reticulum in one locality in a cell. It is believed that certain materials produced within the cell are packaged into membranes, in some cases for export from the cell as secretion granules or to be retained within the cell as lysosomes (see Figure 1.3). Secretion granules accumulate near the cell membrane and their membranes are believed to fuse with the cell membrane so that their contents can be released outside the cell. Lysosomes have been likened to cellular 'dustbins'; they seem to confine breakdown enzymes when dispersal through the cell might be destructive. Within a lysosome, superfluous carbohydrates, fats, and other materials are broken down either to materials which are re-utilized or converted to substances which can be oxidized by mitochondria or to waste products excreted from the cell. It has become clear, recently, that lysosomes play an important role in the dynamics of the cell, the maintenance of a proper balance between building up (biosynthesis) and breaking down (degradation).

Chemical components of cells

Proteins

When reviewing the parts of cells and their function, one type of substance has been prominent, namely, the proteins. Though carbohydrates, fats, nucleic acids, and other substances have been mentioned, proteins, in one form or another, appear to dominate the biochemical scene and play some part in every living activity. The synthesis of proteins from amino acid units is achieved by forming peptide bonds (−CO−NH−) between the carboxyl group (−COOH) of one amino acid and the amino group (−NH$_2$) of another. This is brought about *in vivo* (in life) and *in vitro* (literally 'in glass', i.e. in the test tube) by first activating the carboxyl group to provide the energy needed for the combination of the carboxyl and amino groups. Then, secondly, the combination is catalysed, *in vivo,* by specific enzymes. What is remarkable about ribosomes is their ability to perform such combinations in a special order to obtain specific proteins. Enzymes act as catalysts and not only speed up biological reactions but also ensure that the correct product is formed. Nucleic acids can be seen as a similar combination of small units, not amino acids but nucleotides. Examples of nucleotides are:

adenosine monophosphate　　　　　adenosine diposphate (ADP)

and

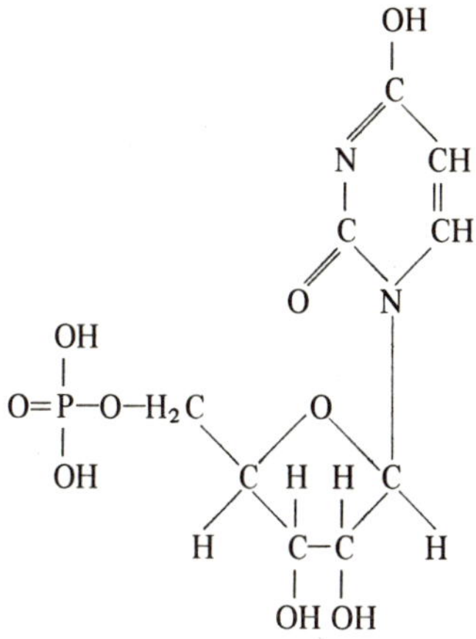

uridine monophosphate

Again, importance lies in the sequence of units joined together and it underlines the extraordinary power of living cells to organize their chemical reactions in a way far superior to anything yet achieved in a test tube.

Carbohydrates

Compared with the precision of protein and nucleic acid synthesis, that of carbohydrates and fats is a less exacting process, and the products are more heterogeneous. A typical, simple sugar such as glucose forms more complex carbohydrates by joining its 1-hydroxyl group to hydroxyl groups of glucose units or other sugars.

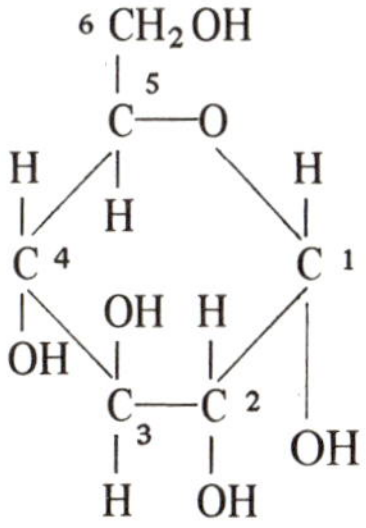

β-D-glucose

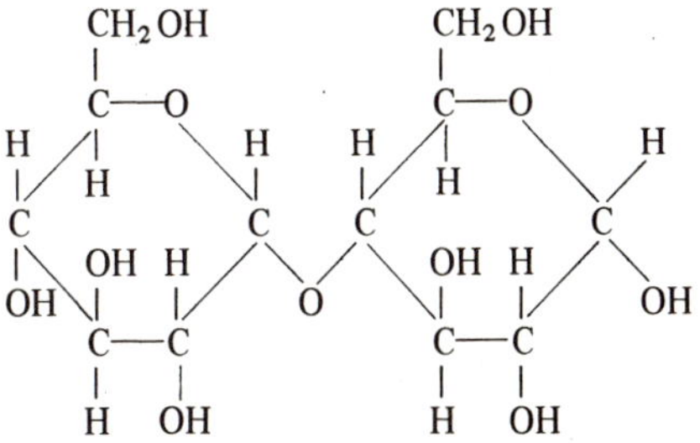

maltose (two glucose units joined 1-4) the α linkage

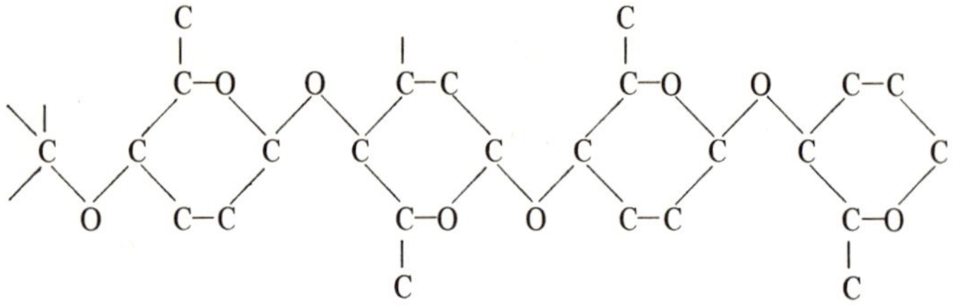

part of a cellulose chain to show β linkage

In milk sugar (lactose), for example, each glucose molecule is linked through its 1-position to the 4-position of a galactose molecule; the linkage is called glycosidic. Although there would seem to be many possibilities for glycosidic combinations between two sugar molecules, the actual kinds of linkages which occur are few in number. In amylose, one of the two components of starch, the glycosidic linkages are formed between the 1 and the 4 positions of adjacent glucose units; there are two possible arrangements of the glycosidic bond around the 1-position of a sugar, these are called alpha or beta linkages. In amylose the linkages are all alpha. A branched chain of glucose units showing 1-6 linkage and 1-4 linkage as found in amylopectin is shown here:

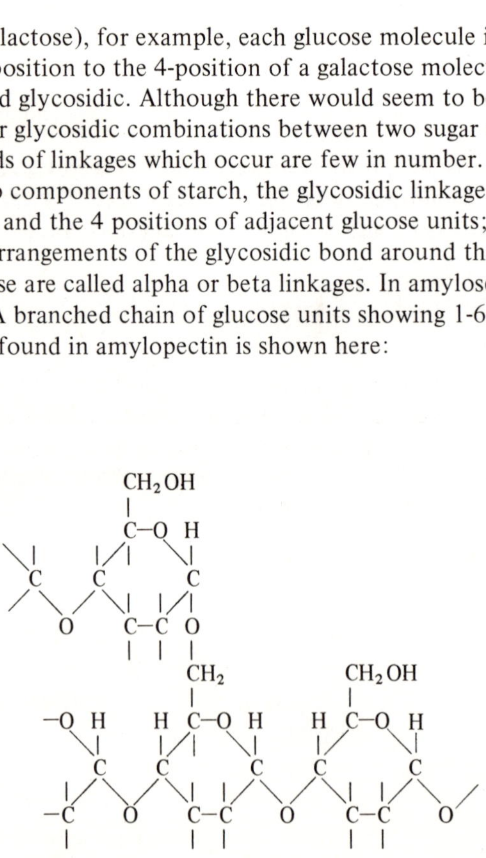

and an amylose chain showing α linkage 1-4 is as follows:

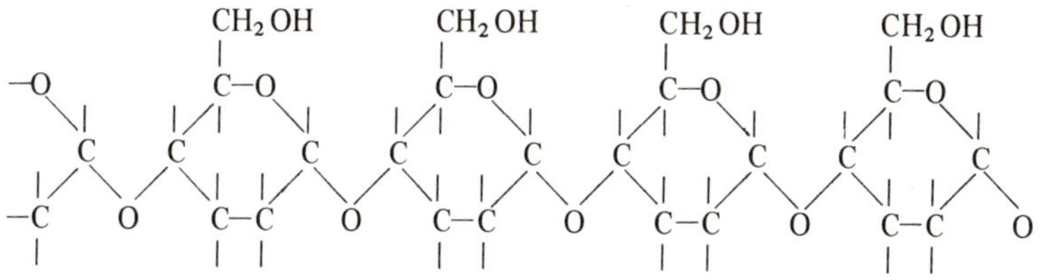

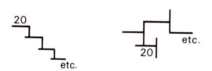

In cellulose the linkages are also 1-4 but beta, and the physical properties of cellulose are quite different from those of amylose due to this apparently trivial difference. Because all the units and links in any one chain are the same it is difficult to see how length is controlled, but we know that a fairly constant average length is maintained by some means (see Figure 1.7).

Figure 1.7. Amylose chains are about 300 glucose units long. Amylopectin can be represented diagrammatically by lines 20 units long with interchain links

Lipids

Simple fats or lipids are esters of glycerol. An ester is the product of combination between an alcohol and an acid. Glycerol is a trihydric alcohol and three fatty acid molecules can join one of glycerol to make glycerol ester or triglyceride. Fatty acids vary both in the number of carbon atoms in the side chain and the possession of double bonds. Common fatty acids have even numbers of carbon atoms between sixteen and twenty-two; some, like stearic acid, have no double bonds, but others, like oleic, have one, linoleic has two, or even three, as is

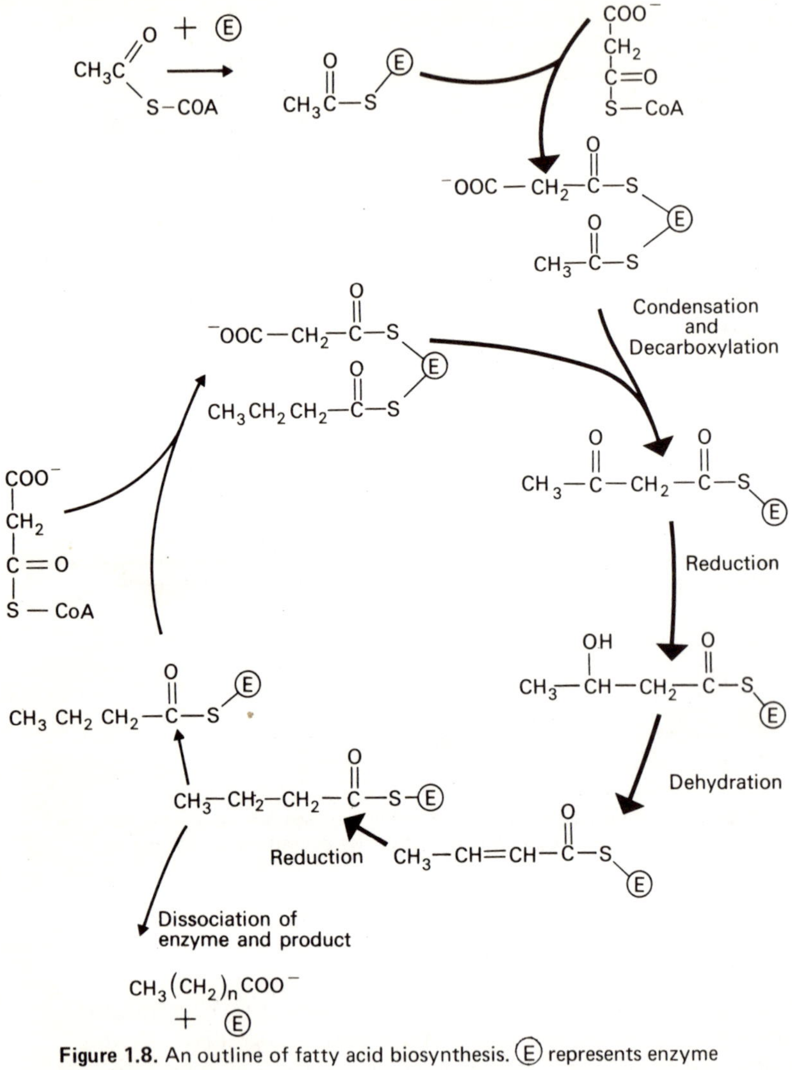

Figure 1.8. An outline of fatty acid biosynthesis. (E) represents enzyme complex

known in the case of linolenic acid. Many combinations are presumed to occur in triglycerides, and different types of fats are found in different kinds of organism and in different parts of the same organism. The molecular components of a fat determine its melting point and it is interesting to note that the melting point of the fats in cold blooded animals is lower than the melting points of those of warm blooded ones. It is lower, too, in marine animals than in terrestrial ones and lower in tissue at the extremities of animals than in the central parts of their bodies.

Proteins, nucleic acids, carbohydrates, and fats are all relatively complex molecules and may have molecular weights of more than one million. These are sometimes called biological macromolecules. They are composed of simpler units—amino acids, sugars, fatty acids—which are themselves synthesized within cells from even smaller, simpler chemical units or molecules. For example, fatty acids are built up from two-carbon acetate units by condensations and reductions (see Figure 1.8). Catalysis of the reaction is done by fatty acid synthetase which is not a single enzyme but a collection of several which governs the reactions. This is an example of a multi-enzyme complex with a molecular weight of 2 000 000; it is so large that its hexagonal outline can be seen under the electron microscope (Figure 1.9).

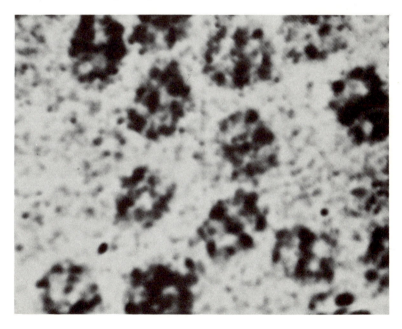

Figure 1.9. Part of an electron micrograph showing molecules of the enzyme fatty acid synthetase (after Hagen and Hofschneider)

The number of enzymes in a living cell is believed to run into tens of thousands. It is difficult to imagine so great a complexity in such a minute volume but it gives some impression of the biochemical sophistication of a living cell. Seventy years of biochemistry have only just begun to unravel the cell as a chemical unit and to discover the interrelationships between its chemical pathways, but a start has been made.

2.

Atoms and Molecules

One of the most characteristic features of life is change. Not only do some living things change their positions, all organisms whether moving or stationary, change in substance. The material of which they are made is constantly changing; living matter perpetually undergoes chemical change To understand how life maintains itself we must try to understand how these chemical changes come about. Reactions are not confined to living organisms. Rocks are eroded by wind and sea, and the rain dissolves and changes the face of the earth, but these changes are slow in comparison with those in living creatures. Because the speeds of reactions are affected by enzymes in living systems, these substances are of enormous importance.

Bonds between atoms

Substances are composed of atoms and only rarely do these exist separately; almost always they are combined together in pairs or larger molecules. The fundamental questions of chemistry include 'what causes atoms to combine as they do?', and 'what causes them to dissociate and recombine in new arrangements?' A chemical reaction is essentially a rearrangement of atoms.

When two or more atoms remain together they are said to be held by chemical bonds, but this does not explain that there are several different ways in which the attachment occurs. There is a tendency of all particles to attract one another, and though the forces of attraction are very small compared with the others to be described, they nevertheless play a part in the chemistry of living organisms. These forces were first recognized

by J. D. van der Waals, investigating the behaviour of gases, and so his
name is given to them.

Covalent bonds

Other bonds depend on the structure of the atoms and the most
common in living systems is the covalent bond in which electron orbitals
of neighbouring atoms overlap. Each atom has a nucleus bearing a
positive charge and negative electrons in orbit round it. Hydrogen atoms
have one of each and others have nuclei with more charges and
consequently more electrons in orbit. The important point to grasp is
that these electrons are grouped in orbitals and each orbital never
contains more than a certain fixed number of electrons; when the
orbitals of an atom are full it is exceedingly stable or inert.

The orbital nearest the nucleus never contains more than two
electrons and the next never more than eight. Helium has two electrons
in its one orbital, neon has two plus eight and both gases are extremely
stable and unreactive; so is argon with two plus eight plus eight electrons
in three orbitals. The more interesting and reactive elements with
intermediate numbers of electrons are capable of forming compounds.
For example hydrogen has one incomplete orbital containing one
electron, carbon has two orbitals, one complete and the other incomplete (two plus four), and chlorine has three orbitals (two plus eight plus
seven electrons). Such elements can form stable compounds by sharing
electrons and making up complete orbitals.

The simple organic compound methyl chloride, for example, consists
of molecules in which three hydrogen atoms and one of chlorine
combine with a single carbon atom. Each bond consists of two shared
electrons, one from each atom. Chlorine has its own seven electrons of
the outer orbital and a share of the eighth so bringing it up to an
argon-like arrangement. Similarly carbon becomes neon-like and
hydrogen like helium. Consequently the compound is more stable than
the component elements in isolation. This arrangement is illustrated in
Figure 2.1, but it must be remembered that, although the electrons are
distinctively marked in the diagram, in fact all electrons are indistin-
guishable from each other. Such diagrams are over-simplifications;
electrons of overlapping orbitals pair up to make regions of 'electron
density' common to the combining atoms. Such bonds of shared
electrons are termed covalent and are much stronger than the van der
Waals forces described above and indeed stronger than the other bonds
to be described later.

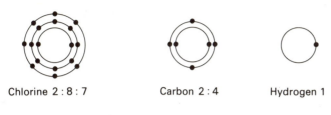

Chlorine 2 : 8 : 7   Carbon 2 : 4   Hydrogen 1

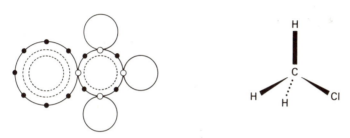

Figure 2.1. Methyl chloride molecule

Bond energy

All compounds can be broken down as the bonds between their constituent atoms are severed, but sometimes these are broken only in extreme conditions of high temperature. All atoms and molecules (above absolute zero temperature) are in motion and so possess kinetic energy. This we can calculate and it has been found to be quite inadequate to account for the energy changes which take place when chemical reactions occur. The energy which seems to reside in chemical substances and which may be partly or wholly released in a reaction when two atoms come together to form a covalent bond is called potential energy. The potential energy of the system (both atoms) decreases due to the overlapping of orbitals. So to break the bond, energy has to be put into the system. A strong bond is one which is difficult to break; it requires more energy for its disruption than a weak bond, and molecules with strong bonds are more stable than those with weaker bonds.

The energy required to separate the atoms united by a bond (measured in standard conditions) is called bond energy, and this depends not only upon the two atoms directly involved but also on atoms in their immediate neighbourhood. This means that the bond energy of, for example, C—H can vary from one molecule to another, and the same bond may alter in bond energy if other parts of the molecule change in composition (see Table 2.1). In biochemical reactions, where energy changes are usually small, uncertainties about individual bond energies are often greater than the overall energy change in the reaction under consideration. For this reason, such energy values are seldom used.

Table 2.1. A comparison of standard free energies of formation of methanol, ethanol, methane, and ethane

	Number of bonds per molecule				
Formula	C–O	O–H	C–H	C–C	Free energy
CH_3OH	1	1	3	0	−205 kJ
C_2H_5OH	1	1	5	1	−235
difference			2	1	+30
CH_4	0	0	4	0	−50
C_2H_6	0	0	6	1	−34
difference			2	1	−16

Note that the contributions made by the two C—H bonds and the single C—C bond differ by 46 kJ in these two kinds of compound.

Electrovalent bonds

Stable molecules can also be formed from atoms which donate electrons, so reducing their number to make the outer orbital like that of an inert gas atom. Sodium for example can donate an electron in the presence of chlorine so that both atoms are stable (2,8,1 donates one electron to 2,8,7). This of course upsets the balance of electrical charges between atomic nuclei and their surrounding electrons. Sodium becomes positively charged and chlorine negatively charged (Na^+ and Cl^-). The process is known as ionization and the bond is simply the attraction between ions of opposite charge. Sodium chloride exists as an orderly arrangement of sodium and chloride ions in equal numbers forming crystals. The electrostatic forces break down when the compound is put in water and a solution is formed. Water has a high dielectric constant*

* The force between two electric charges is affected not only by their strengths (q_1 and q_2) and the distance between them (d) but also by the medium which separates them:

$$\text{force} = \frac{q_1 q_2}{d^2 k}$$

For a vacuum this constant k is unity, for paraffin wax it is 2, for ethanol 26, and for water 81.

which means that the electrostatic forces between ions are dramatically
reduced (approximately eightyfold). Water itself is a polar compound —
its molecules are attracted to either positive or negative charges, so that
hydrated ions form in aqueous solution. The type of bond between
atoms depends on *both* types of atom so that chlorine can bond
covalently with carbon but as an ion with sodium. A variety of bonds
may be found in single molecules of more complex substances. For
example, sodium acetate

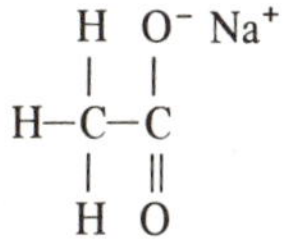

$$
\begin{array}{cc}
\text{H} & \text{O}^-\ \text{Na}^+ \\
| & | \\
\text{H}-\text{C}-\text{C} \\
| & \| \\
\text{H} & \text{O}
\end{array}
$$

one carbon atom is bound to three hydrogens and its neighbouring
carbon covalently but this latter shares one electron with an oxygen
atom (2,6) which has received an electron from sodium thus rendering it
a negative ion. The remaining oxygen is bound to the carbon atom by
sharing *two* electrons, and this is indicated by two lines and called a
double bond. Triple bonds also exist but are much rarer. A further device
for achieving completed electron orbitals is exemplified by benzene,
C_6H_6

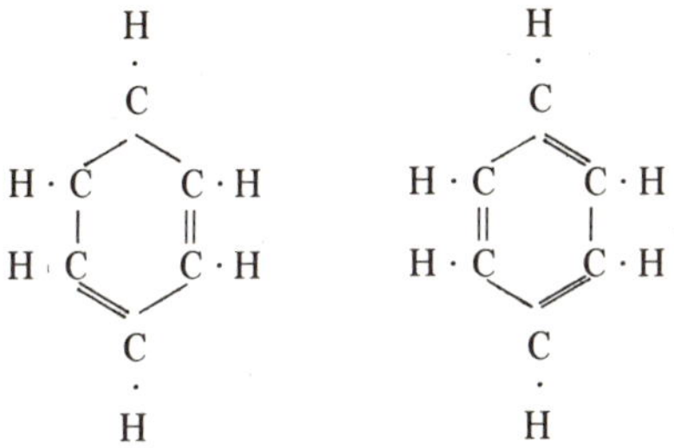

Structural formula    Conventional representation

The molecule is in the form of a ring, and double bonds alternate as
shown in the diagram. Such an arrangement of alternating double bonds
increases the stability of the molecule as a whole. The presence of the
ring structure is widespread in organic compounds so much so that they
are given the special name of 'aromatic' compounds (due to the early
notion that they all had pleasant or pronounced odours). In molecules
such as benzene the ring is planar; the alternating double bonds combine

together so that it is impossible to define just which bond is double or single. This is a case of resonance. It confers extra stability on the aromatic material and results in electron concentration above and below the plane of the ring. The amino acids phenylalanine, tyrosine, and tryptophan, and the nucleic acid bases, derivatives of the purine and pyrimidine ring systems, are examples of aromatic compounds important in living creatures:

phenylalanine   tyrosine   tryptophan

The rare triple bond is an addition of another bond of the second type. It is interesting to note that the bond energies for carbon–carbon are not simply related but are C–C 348 kJ mol^{-1} (at standard temperature) C=C 614 kJ mol^{-1} and C≡C 839 kJ mol^{-1}.

Another kind of electron sharing can take place when a pair from one atom unites with another lacking in electrons. This type of bond is termed 'co-ordinate'. Covalent and ionic bonds play the major roles in holding atoms together to form the molecules of living matter.

Hydrogen bonds

There are also other weaker forces which are important. These are of little concern in small molecules but in the very large ones common in cells a multiplicity of small forces can assume great significance. The first such weak interaction to be recognized was the hydrogen bond. These bonds arise from unequal sharing of electrons. An example is provided by an oxygen atom of acetic acid. Here there is a tendency for the hydrogen atom attached to the oxygen to separate, as an ion, from the rest of the molecule, which is then an acetate ion.

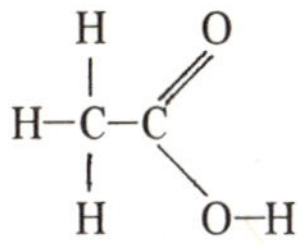

H   O
| //
H–C–C
| \
H   O–H

acetic acid

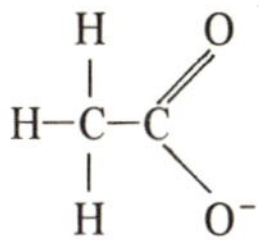

H   O
| //
H–C–C
| \
H   O⁻

acetate ion

The tendency to lose hydrogen atoms, as hydrogen ions, is a characteristic of acids. Acetic acid is a weak acid; the tendency is only slight. The other oxygen atom attached by a double bond also takes more than its fair share of electrons. This is represented by an arrow

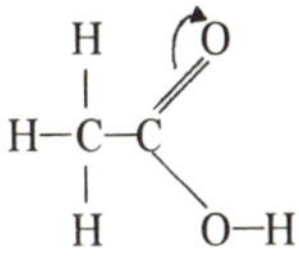

H   O
| //
H–C–C
| \
H   O–H

inequality of electron distribution in acetic acid

it gives rise to a small negative charge on the oxygen atom and a small positive charge on the central carbon atom.

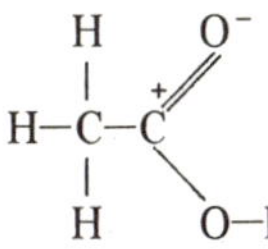

H   O⁻
| //
H–C–C
| \
H   O–H

charge distribution in acetic acid

The bond is said to be polarized and this inequality of charge can result in additional interaction in molecules. It is not confined to C–O bonds but can affect other elements, chiefly of negatively charged participants, nitrogen and oxygen.

A specific interaction can occur between the double bonded oxygen of acetic acid and a hydrogen atom of another acetic acid molecule.

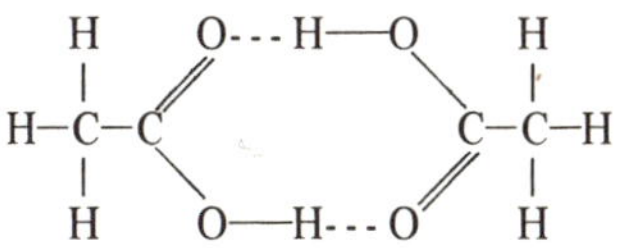

H   O---H—O   H
| //   \ |
H–C–C   C–C–H
| \   // |
H   O—H---O   H

hydrogen bonding between two acetic acid molecules

Not only has the oxygen atom an overall negative charge but the hydrogen atom has less than a fair share of the bond joining it to oxygen. The two identical interactions represented by dotted lines are hydrogen bonds, and because of these, acetic acid dissolved in non-aqueous solvent such as benzene, has a molecular weight of 120, which is double that implied. When acetic acid is dissolved in water it exists, for the most part, as single molecules. Water is itself hydrogen bonded; each water molecule is attached to another by a single hydrogen bond:

On average, each water molecule is hydrogen bonded to one and a half other water molecules, in liquid water. This intermolecular hydrogen bonding accounts for the very high melting and boiling point of this substance (compared, for example, to hydrogen sulphide) and for its high latent heat of vaporization (45 kJ mol^{-1}), one of the highest known. A molecule like acetic acid will form a hydrogen bond with water simply because in a dilute solution of the acid there are more surrounding water molecules with which to interact. So the acid will exist as a hydrated monomer or single molecule. In dilute aqueous solutions, hydrogen bonding occurs preferentially with the solvent. A hydrogen bond is a weak bond formed only between hydrogen atoms and those of nitrogen, oxygen, and fluorine. It is an example of polar interaction; a bond between two atoms which, as a result of bond polarization, has small, opposite electrical charges.

Molecular environment

The molecular structure of a compound and consequently the way it reacts may differ according to the kind of solvent in which it is dissolved. Solvents, like all liquids, can be pictured as a vast multitude of molecules all free to move about at random, confined only by a boundary surface. But this picture is too simple a model because liquids such as benzene and ether are not miscible with water. Their molecules are not free to move at random in any direction. Affinities between different kinds of molecule differ; there are greater affinities, for example, between one benzene molecule and another than exist between a benzene and a water molecule. In water the intermolecular forces concerned are largely hydrogen bonds, but in liquids such as benzene there is insufficient polarization to produce these. There are larger van der Waals interactions between benzene molecules than between benzene and water.

Compounds can be categorized as immiscible or 'water fearing'—given the name hydrophobic—or as 'water loving'—called hydrophilic. Benzene and petrol go into the first, and sodium chloride into the second category. Different forms of the same compound can fall into different categories. For example, some amino acids exist as a doubly ionic form (zwitterion), and this form is more hydrophilic and therefore more soluble in water than the un-ionized form of the same amino acid:

$$\overset{+}{N}H_3-CH_2-CH_2-CH_2-CH\overset{\displaystyle NH_2}{\underset{\displaystyle COO^-}{<}}$$

Zwitterion of ornithine

In large molecules found in biological systems, both hydrophilic and hydrophobic groupings may be found within the same molecule, and the clustering of hydrophobic groups together, brought about by van der Waals forces and the exclusion of hydrogen bonded water in the locality of such clusters, produces what is called a hydrophobic bond.

Within these hydrophobic regions other bonds, unstable in the presence of water, can form satisfactorily. Such molecules can thus behave in an exceedingly complex manner. This illustrates the inadequacy of picturing giant molecules as merely an assemblage of smaller ones.

Giant molecules

For many reasons, large molecules may be able to react in quite a different way from their component sub-units. This is an important concept when considering the reactions of enzymes and other substances commonly found in living systems. Perhaps the best known of these, consisting of very large molecules, is deoxyribose nucleic acid (DNA) (Figure 2.2) already mentioned as a constituent of cell nuclei and carrier of genetic information. Its molecules consist of (1) a pair of deoxyribose-phosphate chains with negatively charged groups (2) pairs of bases, stacked in an offset manner which results in the chains twisting,

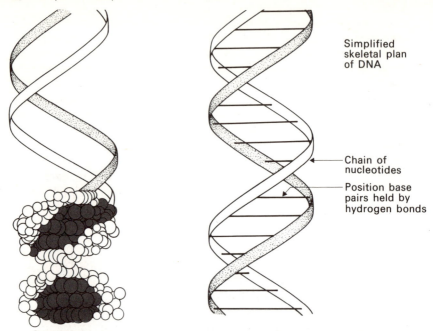

Figure 2.2. Diagram of a model of part of a double helix of deoxyribose nucleic acid (DNA)

and (3) specific hydrogen bonds between base pairs. These last are essential to the theoretical structure, yet it might be argued that in the aqueous situation of living cells hydrogen bonds could not remain stable. It is now believed that interactions occur between the bases stacked almost vertically above each other. The electrons in each base system can interact by van der Waals forces to form a hydrophobic region in the middle of the molecule. These base–base interactions or base stacking forces form a region in which the interbase hydrogen bonds can be stable like the dimer of acetic acid in benzene.

The slight offsetting of the bases, one above the other, is due to the sugar–phosphate chain which constrains the bases into a helical rather than a cylindrical array. It is likely that the deoxyribose–phosphate chain with its many hydrophilic groupings ensures that DNA is miscible with water, a necessary property for a biologically active ingredient of living material. The DNA molecules are so large that they possess two opposing tendencies which coexist within the same molecule. The giant molecule has a hydrophilic exterior, stabilizing the system in its aqueous environment of the cell, and a hydrophobic interior in which important interactions can take place which would otherwise be unstable in such an environment. DNA is not alone in this duality; it is also a feature of protein and fat molecules. Such a dual character is especially important in membrane structure, where transport of predominantly hydrophilic materials through a water-impermeable structure necessitates an essentially hydrophobic (lipid) barrier containing hydrophilic 'holes'— believed to be mainly protein.

3.

Enzymes, Equilibria, and Energy

So far as we know, nothing goes on in living matter which cannot ultimately be explained by physics and chemistry. But, even in the simplest micro-organism, metabolism is such a maze of interrelated reactions, involving large and complex molecules, that we are still far from achieving a complete account of it in these terms. Many of the complexities of metabolic reactions stem from the enzymes which catalyse them.

Suppose that an imaginary substance represented by the formula AB dissociates into A and B reversibly so that the reactions (AB to A plus B) and (A plus B to AB) both occur. Starting with AB alone, dissociation will take place and the products A and B will begin to accumulate so the reverse reaction then begins. Eventually forward and back reactions will go on at the same rate, that is, the amount of change in one direction will be exactly balanced by the amount of change taking place in the opposite direction. At this point the reactions are said to be in equilibrium. In some reversible reactions equilibrium occurs when the concentration of the products is small compared with that of the reactants; in others the reverse is true.

Enzymes do not alter the position of equilibrium but they facilitate reaction in both directions so that the attainment of equilibrium is hastened. These reactions occur only if the energy changes associated with them are favourable. The meaning and significance of this statement needs to be explained and this can be done conveniently by taking one reaction as an example. Hydrogen peroxide is a compound generated in a number of living processes and this can serve to illustrate the point. One such process is the oxidation of phenylalanine to tyrosine. In plant foodstuffs there is more phenylalanine than most animals require. This amino acid is produced from the food during digestion and harmful accumulation is prevented by its oxidation:

phenylalanine   tyrosine

Tyrosine is subsequently broken down to smaller molecules which supply the animal with energy. But if the hydrogen peroxide were allowed to accumulate, it too would upset metabolism, and there is an enzyme, catalase, which brings about its decomposition:

$$2H_2O_2 \xrightarrow{\text{catalase}} 2H_2O + O_2$$

This simple reaction is an example of a chemical change. We know that if liquid hydrogen peroxide is decomposed at 25 °C, 119 000 J mol^{-1} of energy are released. The reaction is not, of course, spontaneous; if it were, we should not know the compound as a liquid bought and sold in bottles, it would merely be a transient, theoretical compound.

Activation energy

Between hydrogen peroxide (the reactant) and water plus oxygen (the products) there is said to be an energy barrier—the 'activation energy'. In a chemical reaction the molecules involved must surmount this barrier. In other words, they remain as they are unless the molecules are supplied in some way or another with a definite amount of energy. Once this is supplied, reaction goes ahead. The presence of enzyme molecules reduces this necessary amount of energy. For hydrogen peroxide the barrier is 75 500 J mol^{-1} but the enzyme catalase reduces this to 8400 J mol^{-1}, a more readily available amount of energy at normal temperatures, so the peroxide decomposes. These energy values are represented diagrammatically in Figure 3.1; note that the energy involved in surmounting the barrier is subsequently recovered. There is no direct connection between the amount of activation energy and the overall energy change. The

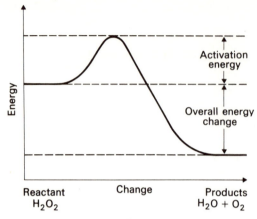

Figure 3.1. Energy values

commonest and simplest way of increasing the energy of reacting molecules so that they surmount the energy barrier is by heating. Once a reaction gets under way it may generate enough energy to sustain itself. An everyday example of this is provided by lighting the gas. Energy supplied by a match overcomes the barrier but once ignited the burning gas maintains the reaction. Organisms operate at fairly constant temperatures, and sudden, local applications of heat are not part of their internal regime; other alternative methods of initiating reactions have evolved.

Free energy

The overall energy change occurring in a reaction is known as the free energy change, symbolized by ΔF (or sometimes by ΔG). In the case of the decomposition of hydrogen peroxide, the free energy change represents a loss to the reaction and is therefore considered negative:

$$H_2O_2 \xrightarrow{\text{catalase}} H_2O + \tfrac{1}{2}O_2 \qquad \Delta F - 119.6 \text{ kJ}$$

In understanding chemical change a great deal of importance is attached to ΔF. If a reaction is to proceed the free energy change must be negative (the free energy of the system must decrease). This is what was meant earlier by the phrase 'when the energy changes are favourable'.

In chemical reactions energy may be derived from two sources, bond energy and entropy. Bond energy is also known as internal energy change (ΔE) and results from the rearrangement of chemical bonds (see Chapter 2). We can think of the reaction outlined above as being a number of component parts as in Table 3.1.

Table 3.1

Reactant(s)	Product(s)	Internal energy changes
H_2O_2 (gas)	2HO	ΔE_1 + 212 kJ mol^{-1}
HO	H + O	ΔE_2 + 430 kJ mol^{-1}
O + O	O_2	ΔE_3 − 495 kJ mol^{-1}
H + O	HO	ΔE_4 − 430 kJ mol^{-1}
H + HO	H_2O (gas)	ΔE_5 − 495 kJ mol^{-1}

In each component reaction only one chemical bond is broken. (The ΔE values of particular bonds are known as bond dissociation energies.) The choice of the sequence of reactions in Table 3.1 is governed by the availability of those bond dissociation energy measurements which are accessible to direct experimental determination. The overall ΔE value of the decomposition of hydrogen peroxide is provided by

$$\Delta E_1 + 2\Delta E_2 + \tfrac{1}{2}\Delta E_3 + \Delta E_4 + \Delta E_5 = -103 \text{ kJ mol}^{-1}$$

This applies to gaseous peroxide and steam; to obtain figures for liquids the gram molecular latent heats of vaporization of peroxide and steam condensation must be taken into account. When this is done the internal energy change becomes −96 kJ mol^{-1}. Oxygen is produced as a gas against the existing atmospheric pressure (the products take up a greater volume than the reactant) and this slightly reduces the ΔE value further to −94 kJ mol^{-1}.

Enthalpy

This final combination of bond energy change and external work is known as the enthalpy change (symbolized by ΔH). Note that if the volume of the whole system is kept constant then no correction for external work is needed; the enthalpy change (at constant volume) is the same as the internal energy change ΔE. Enthalpy changes can be measured by noting temperature changes and calculating the heat liberated. For example, one gram molecule (34 g) of hydrogen peroxide, initially at 25 °C decomposed catalytically in a calorimeter, causes an increase in temperature equivalent to the liberation of 94 kJ mol^{-1} of heat at atmospheric pressure. But the overall energy change expressed in Figure 3.2 is greater (it is more negative) by 25 kJ mol^{-1} than the enthalpy change. Two questions arise—how is this ΔF value known—and, if correct, where does the extra 25 kJ mol^{-1} come from?

Measurements of free energy are not easy and some of the practical difficulties are considered later. Three methods are generally available, through the equilibrium constant as well as by thermal and electrical measurements, the latter two being the more reliable. The reason for a difference between enthalpy change ΔH and free energy change ΔF stems from the way in which ΔH is measured. Usually it is done by measurements of the heat evolved by a reaction. If such measurements could be done without a change in temperature the reaction might be

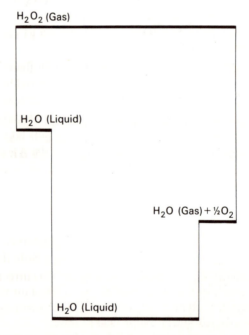

H_2O_2 (Gas)

H_2O (Liquid)

H_2O (Gas) $+$ $\frac{1}{2}O_2$

H_2O (Liquid)

Figure 3.2. Energy change in decomposition of hydrogen peroxide

seen to take heat from its surroundings. Odd though this may sound, it would happen if hydrogen peroxide reaction took place at constant temperature. The missing 25 kJ mol^{-1} would come from the surroundings. This is related to the entropy change ΔS in the reaction. The entropy contribution increases with external temperature T; the two are related thus:

$$\Delta F = \Delta H - T\Delta S$$

So to increase the change in free energy of a reaction, to make it more negative (and thus more likely to take place), an increase in entropy is needed. Conversely, a reduction in the overall energy change in a reaction is associated with a decrease in entropy. When a change takes place which is purely physical such as melting or dissolving, then the necessary energy is derived only from an increase in entropy.

Entropy

The meaning of this word has become rather obscure because a variety of apparently different definitions have been used by various authors depending on their background and approach. For example physicists may refer to entropy as a measure of efficiency of a heat engine. General propositions such as 'the entropy of the universe always increases' tend to be less than helpful to a biologist because they are so general, philosophical, and axiomatic. Biologists study small organisms operating in conditions of approximately constant temperature. In this context one of the most helpful concepts of entropy is that of the degree of molecular disorder. An increase of disorder implies an increase of entropy. A hydrogen peroxide molecule (using the familiar example), like any other at normal temperatures, turns, swings, and vibrates (see Figure 3.3): the rotational, translational, and vibrational energies are

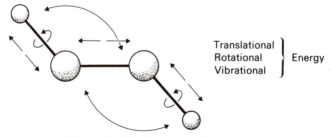

$$\left.\begin{array}{l}\text{Translational}\\\text{Rotational}\\\text{Vibrational}\end{array}\right\}\text{Energy}$$

Figure 3.3. Kinetic energy of a molecule

small compared with chemical bond energies and depend on the prevailing temperature. When decomposition occurs the molecules of water and oxygen have more modes of vibration so that more energy is needed to cause the atoms to rotate and vibrate about one another. The two product molecules have more degrees of freedom than the reactant molecule. This molecule is thus, in a sense, more complex than its products. The energy necessary to sustain the extra vibrational and rotational modes is represented by $T\Delta S$ in the equation. ΔS is a measure of the increase in the number of degrees of freedom and the energy associated with each of these is determined by the external temperature T.

'Unfavourable' reactions

The ΔH and ΔS values contributing to the overall energy change ΔF need not necessarily both cause a decrease in free energy. In some reactions ΔH is positive and measurements show that reaction causes a lowering of temperature; these are called endothermic. The simplest endothermic reactions include the formation of solutions from salts and water. The break up of crystals and the hydration of the released ions leads to an *increase* of bond energy. In spite of this the reaction does take place so we expect there to be an overall decrease of free energy, and this is accounted for by an increase of entropy (the change is positive) greater than enthalpy (positive enthalpy − positive entropy = negative free energy). Many reactions are associated with negative values of entropy (a decrease in disorder) as for example the synthesis of ammonia:

$$N_2 + 3H_2 \rightarrow 2NH_3$$
$$\tfrac{1}{2}N_2 + \tfrac{3}{2}H_2 \rightarrow NH_3$$

For every gram molecule of ammonia formed at 25 °C, −16.8 kJ is released. The enthalpy change as measured experimentally is −46.2 kJ mol^{-1}. The difference of −29.4 kJ mol^{-1} is due to a decrease in entropy. There are fewer degrees of freedom in an ammonia molecule than in those of the constituent elements. The enthalpy value comes from bond energy present in hydrogen and nitrogen but not present in the ammonia molecule. It may seem odd that ammonia synthesis apparently provides energy at the comparatively low temperature of 25 °C. In fact, even though the free energy is favourable for change the activation energy is not. Nitrogen molecules are unreactive 25 °C so to get the reaction started there must either be a catalyst present to reduce the activation energy requirement or heat must be supplied. The latter is not a productive condition because by raising the temperature T, the value of $T\Delta S$ is raised and as ΔS is negative this time, then as the temperature is raised so ΔF is rendered positive and the reaction becomes energetically unfavourable. In industry, ammonia is made by some raising of temperature and the use of a catalyst, but some algae and bacteria possess a catalyst (enzyme) system capable of producing ammonia at ordinary temperatures. The enzyme is nitrogenase capable of activating nitrogen and thus 'fixing' it. Before man began making nitrogen fertilizers for the soil the nitrogenase was probably the only means by which nitrogen of the air became part of the metabolism of plants and animals.

Most reactions in living creatures occur in dilute solution and the synthesis of ammonia is no exception. In these circumstances ΔF (solution) is -27 kJ mol^{-1} when the ammonia is considered to be present as a molar solution in water. Neither hydrogen nor nitrogen is particularly soluble in water so that the difference between ΔF and ΔF (solution) is all made up virtually from the dissolution of ammonia in water. A 1 M solution is considerably more concentrated than a gas at atmospheric pressure where the concentration is equivalent to 4.4×10^{-2} M. Also, substances which dissolve in water combine chemically with it; they become hydrated. Such chemical changes have free energy changes associated with them and these changes are additive. So we must consider two values:

$$\tfrac{1}{2}N_2 + \tfrac{3}{2}H_2 \rightarrow NH_3 \text{ (gas) } \Delta F_1 - 16.8 \text{ kJ mol}^{-1}$$

$$NH_3 \text{ (gas)} \rightarrow NH_3 \text{ (solution) } \Delta F_2 - 10.1 \text{ kJ mol}^{-1}$$

$$\text{so } \tfrac{1}{2}N_2 + \tfrac{3}{2}H_2 \rightarrow NH_3 \text{ (solution) } \Delta F_3 (\Delta F_1 + \Delta F_2) - 26.9 \text{ kJ mol}^{-1}$$

The additive nature of bond energies has been assumed in the case of hydrogen peroxide decomposition, above, and similar considerations apply to free energy changes.

Energy fixation

The chemistry of living things is complex, involving a large number of different reactions. Ultimately all reaction would cease if it were not for the sun, from which the energy for all living processes comes. Light from the sun is 'fixed' by green plants during photosynthesis and this is made available to animals, as food, mainly in the form of carbohydrates and oils. One of the most active plant producers of carbohydrates is sugar cane and the sucrose it produces can be regarded as, and has been called, 'locked-up sunshine'. This sucrose is broken down during respiration to provide energy. Before proceeding further we should ask: how is this energy actually stored in the sugar molecules? Is it something to do with the fact that sugar is more complex than the carbon dioxide and water from which it is derived? Do the chemical bonds in sucrose molecules differ from those in carbon dioxide and water or are there just more such bonds? Some insight may be gained by looking at the energy changes involved.

The oxidation of sucrose can be summarized by the equation:

$$C_{12}H_{22}O_{11} + 12O_2 \rightarrow 12CO_2 + 11H_2O$$

The free energy is -5863 kJ mol^{-1} and the enthalpy contribution is -5670 kJ mol^{-1} so the $T\Delta S$ term contributes a mere 193 kJ mol^{-1}, or just over 3 per cent of the total energy change. As the entropy change in a reaction is a measure of the complexity of the starting substances compared with their products, we can see that the apparent complexity of sucrose does not play a significant role.

What of the bonds in sucrose molecules? The internal energy change in this reaction is the same as the enthalpy change since no gas volume change occurs at 25 °C. If we add the values of bond energies (derived from other substances) of the ten C–C, fourteen C–O, eight O–H and fourteen C–H bonds, no energetic abnormalities arise; the total agrees with expectation and if we look at the *number* of bonds as a whole (46) there is no change in total number when oxidation produces carbon dioxide and water ($12CO_2 + 11H_2O$; $12 \times 2 + 11 \times 2 = 46$).

The so called 'locked-up' energy is determined by the *types* of bond present at the beginning of the reaction compared to those at the end. Note that in sucrose there are C–C and C–H bonds, whereas in the products there are only C–O and O–H bonds. Compounds containing a high proportion of C–C and C–H bonds are found to be good fuels when burned to carbon dioxide and water. Oils are better foods than carbohydrates in the sense that they provide more metabolic energy, weight for weight, than carbohydrates like sucrose. This is due to the very high proportion of C–C and C–H bonds in their molecules. They are a major energy source and energy store of most of the larger animals in which body weight is at a premium.

Enzyme systems

Though some understanding of enzymes can be gained from the study of single enzymes and their substrates, this gives little idea of the way in which the most important ones function in life. Complex chemical operations are carried out in cells by sequences of enzymes functioning together. Multi-enzyme systems work so that the product of one reaction is a reactant in the next. An example of a relatively simple multi-enzyme system is provided by glycolysis, the breakdown or degradation of glucose to simpler substances, a process which leads in some organisms to fermentation and the production of ethanol and carbon dioxide. Glycolysis also occurs in muscle and has been the subject of intensive research for many years. Studies of fermentation marked an unconscious beginning of biochemistry in the late eighteenth century, though it was not until 1878 that the term 'enzyme' was coined as something 'in yeast' with special properties.

The enzyme system responsible for glycolysis is remarkable in that it can be extracted from living cells without complete loss of activity. Naturally early workers, including the brothers, Buchner, assumed that they were dealing with a single enzyme (they called it 'zymase') but later realized that the situation was more complex. There are eleven enzymes in the system, each responsible for a different stage in the process which is shown in Figure 3.4. It is shown as far as the formation of lactate. In life these enzymes are, as is known, freely dispersed in the fluid contents of cells, not attached to organelles (if they were attached then presumably they would not be easily extractable). The breakdown of glucose would be pointless if it were not for the accompanying synthesis of adenosine triphosphate (ATP) from adenosine diphosphate (ADP) and phosphate (see page 10). This is not as simple as it sounds, for glycolysis begins with the phosphorylation of glucose, and this can be done only by reaction with ATP which is itself a product of glycolysis! Needless to say, more ATP is produced than is used and the excess ATP is involved in a number of different energy-requiring reactions in the cell's metabolism. The making of ATP can justifiably be regarded as the purpose of glycolysis. Because ATP can cause many thermodynamically unfavourable reactions to occur, it is a vital participant. Without a continual supply of ATP in the cell, metabolism would very soon cease altogether.

It is interesting to note, in passing, that this supply never comes from outside the cell. The first substrate of enzyme action, glucose, enters cells via the bounding membrane and the ultimate products of respiration (carbon dioxide and ethanol or water) pass out in the same way.

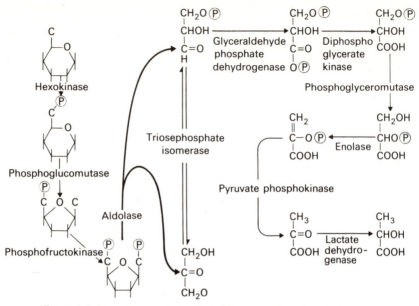

Figure 3.4. The enzymes of glycolysis. P represents a phosphate group

Phosphate and all the intermediates of metabolism remain within the cell undergoing synthesis and degradation inside this limited volume. The products of glycolysis are considered further in Chapter 6.

Biosynthesis

Growth is a characteristic of all living creatures and this implies the biosynthesis of complex polymeric molecules such as nucleic acids, proteins, and carbohydrates. A relatively simple polymer which is found in muscle and liver tissue is glycogen (Figure 3.5). Glycogen is made of glucose molecules condensed together in a very special way. No means is known of directly polymerizing glucose into glycogen *in vitro* or *in vivo*, for the good reason that the free energy change is unfavourable ($+14.7$ kJ mol^{-1} glucose). The glucose has to be activated chemically to provide the necessary energy. This is done by phosphorylation of the glucose by ATP catalysed by the enzyme hexokinase; this reaction has a favourable free energy (-14.7 kJ)

$$\text{glucose} + \text{ATP} \xrightarrow{\text{hexokinase}} \text{glucose-6-phosphate} + \text{ADP}$$

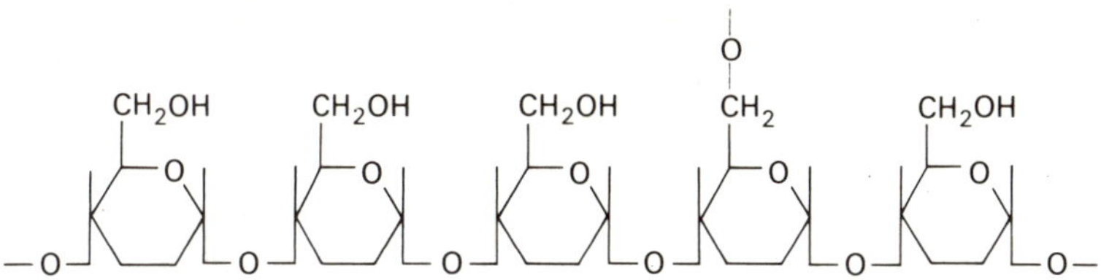

Figure 3.5. The polymer glycogen

Subsequently the glucose-6-phosphate is converted to glucose-1-phosphate and then by a short sequence of reactions involving another nucleoside triphosphate, uridine triphosphate (as outlined in Figure 3.6) is incorporated into glycogen. The overall process has a free energy of $-62.6 \text{ kJ mol}^{-1}$ glucose incorporated into glycogen.

The hexokinase reaction involves ATP; the subsequent biosynthesis of glycogen, uridine triphosphate.

The role of ATP

These triphosphates (particularly ATP) have been found to take part in very many metabolic reactions and this has led to ATP being called the 'energy currency' of the cell because it is said to act rather like money in human society. The ATP is generated in each cell from the ADP in the process of respiration. The most important feature of the metabolism of these triphosphates is their involvement in reactions which would not otherwise proceed at all. Not only are many forms of biosynthesis associated with ATP but also muscle contraction, movements of flagella, cilia, and the transport of ions such as sodium and potassium are dependent on the presence of ATP molecules and appropriate enzymes. In all these processes work is done and energy is provided by ATP. Consequently much attention has been focused on the thermodynamic aspects of ATP chemistry. We can see ATP involvement in terms of hydrolysis and represent it as:

$$\text{ATP} + \text{H}_2\text{O} = \text{ADP} + \text{(P)}$$

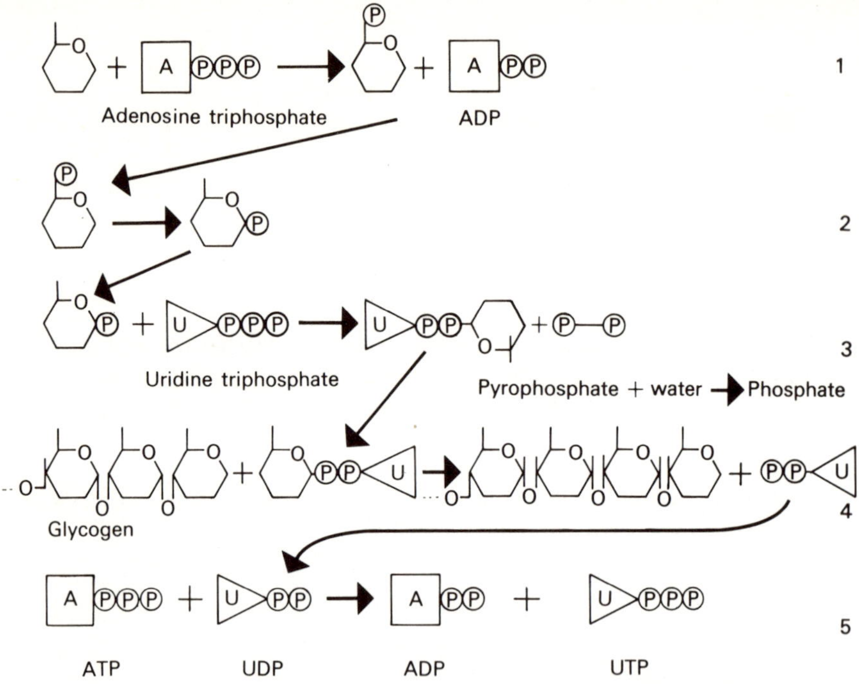

A glucose unit is added to an existing glycogen chain in line 4

Figure 3.6. Series of reactions by which glucose is incorporated into glycogen

and the 'activation' of glucose, referred to above, as:

$$\text{glucose} + (P) \rightarrow \text{glucose-6-phosphate} + H_2O$$

The free energy change of the latter is $+14.7$ kJ mol^{-1} glucose. From this we can suppose that the free energy of ATP hydrolysis is about -29 kJ mol^{-1} since the free energy change in the hexokinase reaction is -14.7 kJ mol^{-1}. There is nothing very remarkable in this. A reaction with a favourable free energy (ATP hydrolysis) is coupled with one having an unfavourable free energy change, such as glucose phosphorylation, to provide an overall favourable transformation. The coupling is achieved through the enzyme hexokinase whose catalytic specificity is such that the two reactions must proceed simultaneously or not at all.

High energy bonds?

It is only when the free energy change associated with the hydrolysis of
ATP is compared with that of other phosphates, such as glucose and
glycerol phosphates, that the high value of ATP hydrolysis becomes
apparent (it is about twice that of the hydrolysis of glucose-6-phosphate).
It is this difference between phosphate hydrolysis free energy which
has caused controversy over the thermodynamic aspects of ATP lasting
almost half a century!

$$\Delta F(\text{kJ mol}^{-1})$$

acetylphosphate	-42.4
adenosine triphosphate	-29.4
glucose-1-phosphate	-21.0
glucose-6-phosphate	-13.8
glycerol-1-phosphate	-9.7

Early measurements indicated that the enthalpy of ATP hydrolysis was
abnormally great; since this usually involves bond energies, the idea arose
that there was something odd about the bonds of ATP molecules and
some were dubbed 'high energy bonds'. Speculations about the three
phosphate groups being locked in a kind of molecular spring were
fashionable for a time but were finally dispelled by the formulation of a
structure of the ATP molecule based on X-ray analysis, see Figure 3.7.

Early measurements of the free energy of hydrolysis of ATP were
based on thermal calculations of the energies involved; these were
inaccurate and proved wrong (see p. 107). More recent and reliable data
have come from determinations of the *equilibrium constant* for the
reaction. Free energy is related to the equilibrium constant—a reaction
with a large, negative free energy change (which means a powerful force
driving it to completion) necessarily has a large equilibrium constant; for
the decomposition of hydrogen peroxide $\Delta F = -119$ kJ and the
equilibrium constant K is about 10^{20}. The precise relationship at
temperature $T\,^{\circ}$ K is

$$\Delta F = -RT \ln K$$

where R is the gas constant measured in kJ $^{\circ} \text{K}^{-1}$ mol^{-1}. The
equilibrium constant for the hydrolysis of ATP is so large (about 10^6)
that it is quite impossible to measure the quantity of ATP in an
equilibrium mixture because there is always an overwhelming excess of
ADP which is chemically very similar to ATP. It is by no means easy to

set up the equilibrium in the first place. This is due to the fact that there is no really suitable enzyme to catalyse the reaction and this, in turn, is because it is a highly undesirable reaction from the point of view of a living organism. Enough has already been said to suggest the value of ATP in cell metabolism. Direct reaction with water to produce ADP would be a waste of respiratory energy. The best means so far devised of achieving the experimental hydrolysis of ATP, for thermodynamic studies, involves two reactions in the formation and hydrolysis of the amino acid glutamine. The formation is catalysed by an enzyme extracted from pea seedlings, glutaminyl transferase.

$$\begin{array}{ccc}
\text{COOH} & & \text{CONH}_2 \\
| & & | \\
\text{CH}_2 + \text{ATP} + \text{NH}_3 \xrightarrow{\text{glutaminyl transferase}} & \text{CH}_2 + \text{ADP} + \text{(P)} \\
| & & | \\
\text{CH}_2 & & \text{CH}_2 \\
| & & | \\
\text{NH}_2-\text{CH}-\text{COOH} & & \text{NH}_2-\text{CH}-\text{COOH} \\
\text{glutamic acid} & & \text{glutamine}
\end{array}$$

The hydrolysis of glutamine is catalysed by a bacterial enzyme gluta-minase, extracted from the gangrene organism *Clostridium welchii*:

$$\text{CONH}_2 + \text{H}_2\text{O} \xrightarrow{\text{glutaminase}} \text{COOH} + \text{NH}_3$$

For both reactions the equilibrium constants, and hence the free energy changes, are capable of being measured directly by the estimation of products and reactants present in the equilibrium mixture. At pH 7, the values are -15.1 kJ mol^{-1} and -14.3 kJ mol^{-1} respectively so, by adding the two, a value for ATP can be derived -29.4 kJ mol^{-1}.

This value is undoubtedly greater than the free energy change associated with the hydrolysis of glucose-6-phosphate and other compounds. At pH 7 the reactants are charged and the equation can be written

$$\text{ATP}^{4-} + \text{H}_2\text{O} \rightarrow \text{ATP}^{3-} + \text{HPO}_4^{2-} + \text{H}^+$$

which shows the generation of a hydrogen ion which tends to favour the hydrolysis reaction. This does not occur in the hydrolysis of glucose or glycerol phosphate. Though pH is an important factor in the hydrolysis of the so called 'high energy' phosphate compounds it must be acknowledged that it is not the only one, and there are other factors concerned both with bond energies and the entropy of hydrolysis, of which we are at present uncertain.

Why phosphates?

Naturally two questions arise. Why is phosphate used by organisms rather than other groups, and what is it about ATP that makes it a participant in so many vital reactions? There are no generally agreed, straightforward answers to these questions, but several ideas have been put forward. It is perhaps the solubility of phosphate and its relative stability at (neutral) pH 7 that make it such an acceptable constituent of cell metabolism. Adenosine may be nothing more nor less than a molecular shape to be 'recognized' by enzymes. Enzymes control the exceedingly complex biochemistry of the cell by their presence or absence. It is just conceivable that phosphate by itself or attached to a very simple organic molecule might be too easily confused and mistaken

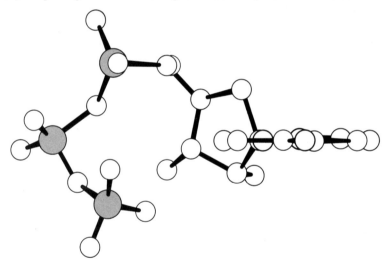

Figure 3.7. Three-dimensional structure of an ATP molecule based on X-ray diffraction studies. The circles represent carbon, oxygen, and phosphorus atoms. Note that the purine ring is on the right with its plane at right angles to that of the page. The ribose ring is in the same plane as the page. The phosphorus atoms are shown stippled. The distance between the terminal phosphate group and the ribose ring is greater than it appears in the figure

by enzymes. This is not to invest enzymes with peculiar powers of perception but it is known that their specificity is acquired through the fitting of substrate and enzyme in an intimate lock and key manner. The suggestion is that ATP needs a complex shape so that it may react only with certain enzymes and thus preserve order in the environment of cellular ultrastructure. It should be noted that ATP possesses a hydrophobic region (the purine ring system) which may be of crucial importance in the initial binding to an enzyme in an aqueous environment, whilst there is also present a highly polar 'tail' (Figure 3.7) which includes the essential phosphate groups. We know that polar reactions are favoured by the aqueous environment around them. Finally, the energy changes involved in these reactions are small and not likely to upset the isothermal state of most living tissues.

These ideas are speculations and doubtless many more will be made. The first task of biochemists is to understand what happens in the living cell rather than why some compounds take part in metabolism and others do not. But it is natural and human to wonder and, in so far as it causes questions to be asked which lead to further investigations, such speculation is profitable.

4.
Enzyme Activity

Enzymes are substances produced in living organisms and they act as
catalysts. They all have one constituent in common—protein. They are
difficult to define more precisely because other constituents vary widely
and include lipids, carbohydrates, and metal ions. They were described
early this century as 'complex nitrogenous substances akin to albumen'.
The ever increasing amount of research done since then has revealed a
wealth of complexity and made precise definition even more difficult to
achieve.

Industrial applications

The intense interest in enzymes is due to several factors. One of these is
the obvious fact that they cause reactions to occur more readily and, as
the chemical industry depends upon reactions, a study of enzymes is
likely to repay the effort. Unlike most non-biological catalysts, enzymes
are highly specific, in many cases catalysing only one kind of chemical
reaction. Studies of cell chemistry have revealed that enzyme activity can
be controlled within the living cell so bringing about co-ordinated
reactions and avoiding the metabolic disorder which would occur if all
possible reactions took place all the time.

The great effectiveness of enzymes as catalysts suggests a potential
asset for industry and they have been used in a number of ways such as
beer clarification (papain) and sugar assay (glucose oxidase and
peroxidase). More recently washing powders containing enzymes have
been promoted and marketed. These contain a bacterial enzyme
(subtilisin from *Bacillus subtilis*) which digests proteins. The idea behind
such powders is that many of the stains on clothing which are difficult to
remove are protein but unfortunately some fabrics such as wool and silk

are also protein so the 'biological action' of detergents must be used with discretion and knowledge.

One of the drawbacks in the use of enzymes for industrial processes is the difficulty in recovering them once reaction has taken place. Unlike many industrial catalysts, enzymes act in a liquid medium. A means of overcoming this difficulty has been the attachment of enzyme molecules to larger particles of inert chemical carriers, mainly synthetic polymers. When these are introduced into a reaction mixture as a fine suspension of polymer beads they can be recovered by filtration after the reaction has taken place. Alternatively, polymer-bound enzymes can be used as a bed through which the reaction mixture is percolated. Because enzymes are made only by organisms their production and extraction is always expensive. Reutilization may well make enzymes an economic proposition in many industrial processes in future.

Intermediate compounds

Whether enzymes are bound to large inert molecules or freely suspended in water, they facilitate reaction by somehow reducing activation energy. Presumably this is done through some interaction with the substrate molecules. If we take, again, the example of hydrogen peroxide decomposition, presumably catalase molecules come into contact with peroxide molecules. We know that in solution catalase molecules are far larger and slower than those of peroxide and that the enzyme's action is specific. Presumably, therefore, we are not concerned with mere molecular collision but with chemical changes involving the formation of bonds even if they are short lived. If some intermediate enzyme–substrate compound does exist, its formation and decomposition will presumably follow the laws of kinetics and in particular the law of mass action (the rate of a chemical reaction is proportional to the concentrations of the reactants). Thus if the concentration of the enzyme catalase is e and that of the substrate s then the rate of formation of the enzyme–substrate complex r_1 is proportional to both:

$$r_1 = k_1 e \times s$$

where k_1 is the rate constant for the reaction. The enzyme–substrate complex is not, of course, the final product of the reaction, it is unstable and can break down in two ways, either (1) to form the starting materials (catalase plus peroxide) with a rate constant k_2, or (2) to form the 'products' catalase, water, and oxygen with a rate constant k_3. If the concentration of enzyme-substrate complex is es then the rate of its decomposition will be r_2

$$r_2 = k_2 es + k_3 es$$

The overall rate of reaction as judged by the appearance of products is $k_3 es$ and in this example of catalase the high overall energy change (118 kJ mol^{-1}) ensures that the reverse reaction does not take place. The enzyme–substrate complex does not accumulate but it presumably exists so it is fair to assume that it occurs in steady state; its rate of formation equals its rate of decomposition. Such an assumption is justified if it leads to reliable prediction, and experiments have confirmed it. Making this assumption we can write

$$r_1 = r_2$$

and

$$k_1 e \times s = (k_2 + k_3)es$$

If the *initial* enzyme concentration, or the total quantity of enzyme, is e_0 then

$$e_0 = e + es$$

. This substitution provides an expression for es in terms of substrate concentration, initial enzyme concentration, and three rate constants

$$es = \frac{k_1 e_0 s}{k_2 + k_3 + k_1 s}$$

and so the overall rate of reaction v is

$$v = \frac{k_3 k_1 e_0 s}{k_2 + k_3 + k_1 s}$$

This is more generally written using the constant K_m and is a combination of all three rate constants

$$K_m = \frac{(k_2 + k_3)}{k_1}$$

so,

$$v = \frac{k_3 e_0 s}{K_m + s}$$

This equation helps in the interpretation of experimental data. If the rate of reaction v is plotted against substrate concentration s with the enzyme concentration kept constant, then a curve as shown in Figure 4.1 is obtained. With a low concentration of substrate, when the value of s is far below that of K_m, then

$$\frac{k_3 e_0 s}{K_m + S} \simeq \frac{k_3 e_0 s}{K_m}$$

and this corresponds to the linear increase of v with s. When the substrate concentration s is much larger than K_m then

$$\frac{k_3 e_0 s}{K_m + s} \quad \text{becomes} \quad \frac{k_3 e_0 s}{s} \quad \text{or simply} \quad k_3 e_0$$

and v becomes independent of s, and the curve becomes a straight line parallel with the abscissa. This value of v is the maximum reaction velocity or v_{max} and the enzyme is said to be saturated because all of it is present as enzyme–substrate complex. Graphs of this kind were first obtained from experiments by Michaelis and Menten in 1913 using the enzyme invertase which catalyses the production of fructose and glucose from sucrose. These workers showed that experimental results matched theoretical expectations and the idea of an enzyme–substrate complex with a limiting, maximum reaction velocity; K_m is called the Michaelis constant.

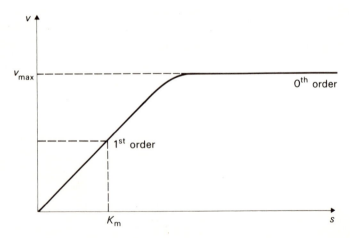

Figure 4.1

This evidence for enzyme–substrate complexes was indirect, and at first no method was known by which they could be demonstrated. In the case of the enzyme chymotrypsin which catalyses the breakdown of proteins the enzyme–substrate complex behaves normally at pH 7. In such neutral conditions it is decomposed as rapidly as it is formed, but at pH 5 breakdown is suppressed and the complex can be recovered. Sometimes an enzyme possesses physical properties which lend themselves to the identification of complexes; catalase is an example. Its molecules contain atoms of iron and the enzyme absorbs green light more than other parts of the spectrum. When added to the substrate, hydrogen peroxide, the wavelength of absorbed light changes—the complex has a different absorption spectrum from that of the pure enzyme.

Use of the Michaelis constant

The Michaelis constant K_m is peculiar to each enzyme-catalysed reaction. Where a single enzyme is capable of catalysing a number of reactions with different substrates, the constant varies with each substrate. For example hexokinase catalyses the transfer of a phosphate group from adenosine triphosphate to glucose molecules and to certain other sugars (Figure 4.2). The Michaelis constants for each reaction differ, the smaller the K_m the faster the rate of reaction. Michaelis thus provided a useful instrument by which experimental data could be interpreted and understood. Actual measurement of K_m from a plot is often difficult because the exact part at which v_{max} is reached is hard to determine. At v_{max}, substrate is used up rapidly so it may not last long

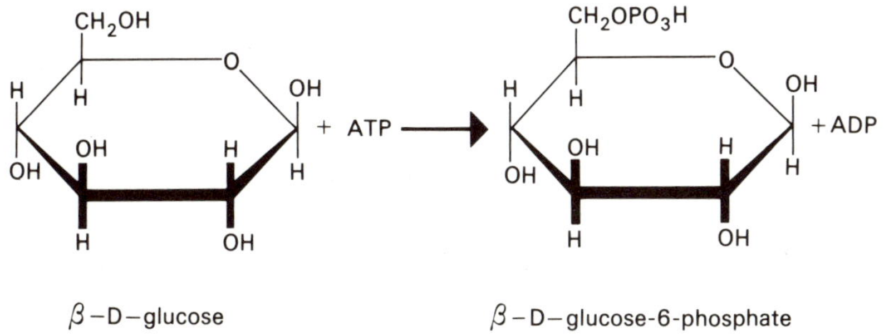

β–D–glucose β–D–glucose-6-phosphate

Figure 4.2

enough for accurate measurements to be made. Alternatively, the relatively large amounts of substrate required may make product and rate measurements difficult. If the equation

$$v = \frac{k_3 e_0 s}{K_m + s}$$

is inverted, it becomes

$$\frac{1}{v} = \frac{K_m}{k_3 e_0 s} + \frac{1}{k_3 e_0} = \frac{K_m}{v_{max} s} + \frac{1}{v_{max}}$$

By plotting $1/v$ against $1/s$ a linear relationship is obtained (see Figure 4.3). The intercepts obtained by $1/v = 0$ or $1/s = 0$ enable K_m and v_{max} to be found easily. Such a graph is named after its originators Lineweaver and Burk.

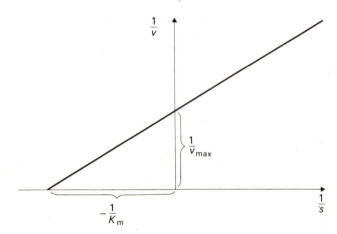

Figure 4.3

The enzyme–substrate complex is the result of chemical combination involving bonding of several kinds including covalent bonding and hydrophobic interactions. The latter can involve whole regions of the enzyme protein. Covalent bonding, on the other hand, is precise and specific and quite often the amino acids serine (hydroxyl group), cysteine (thiol group), and histidine (imidazole group) in the protein molecule are involved, though other amino acid links occur too. The relatively small number of known points of attachment between

substrate and enzyme are not by themselves particularly significant. If
they were, a mixture of a few amino acids would act as an enzyme and
this is not the case. What is also of much greater importance is the
distance and arrangement of these link points in the whole enzyme
molecule. This molecular architecture of some enzyme molecules has
been revealed by X-ray crystallography.

Chymotrypsin

The enzyme chymotrypsin has been studied in great detail in this way
and its chemical action is known. It catalyses the rupture of peptide
bonds in protein molecules but, curiously enough, not all peptide bonds.
Only the peptide bonds associated with naturally occurring L-forms of
phenylalanine, tyrosine, or tryptophan are broken by chymotrypsin (see
Figure 4.4). The breakdown reaction is believed to take place in the
following stages:

(1) binding of chymotrypsin to the substrate which involves
hydrophobic interactions
(2) bonding of a serine unit of the enzyme with the acyl group of
amino acid A1
(3) the rupture of the peptide bond and departure of the amino acid
portion A2
(4) dissociation of the A1-enzyme complex.

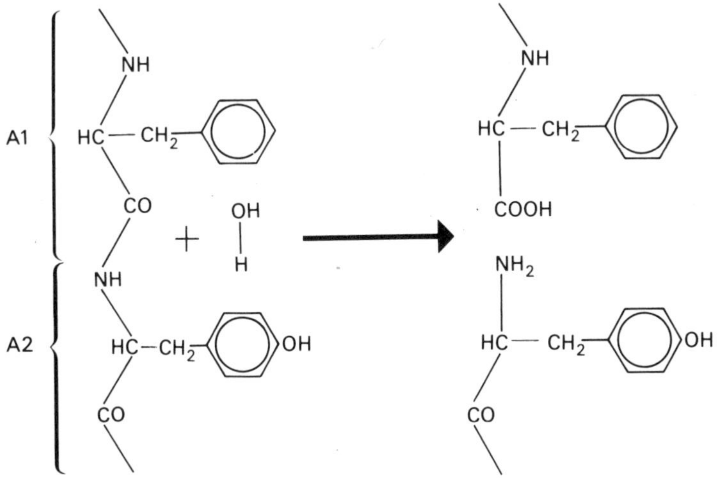

Figure 4.4. Hydrolysis of a peptide bond

Each of these stages has its own activation energy and the highest of these is the last (4) and the most important. If this energy (58 kJ mol^{-1}) is attained then the other stages in the process can also take place. Part of the sequence is represented by Figure 4.5. Of the three points of attachment proposed in Figure 4.5 (i) only one is supported by reliable evidence, a serine hydroxyl group of the 95th amino acid unit in the chymotrypsin molecule (a), (b) is a hydrophobic interaction with the side chain of the amino acid and (c) is probably a polar interaction, which would be unstable in water, but evidence about the structure of chymotrypsin from X-ray crystallography (Figure 4.6) suggests that such interactions could well exist in stable form within the enzyme.

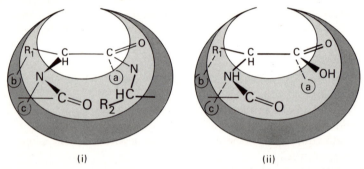

(i) (ii)

Figure 4.5. Diagrammatic representation of points of attachment between enzyme and substrate

The breakdown of the acyl–enzyme intermediate (2) also illustrates another feature of enzyme mechanism, that of optimal placing of reactive groups in enzyme molecules. If an enzyme like chymotrypsin is to function efficiently, the various binding sites must be in those positions most conducive to the complex, multi-stage process outlined above. If, for example, the first bond made is at site (b), then it is easier for sites (a) and (c) to form their attachments to the substrate. In the watery environment of a living cell it may well be that the hydrophobic bond at (b) is the first to form. Such a first step would facilitate reaction at sites (a) and (c). De-acylation of the acyl–enzyme complex is helped by the convenient placing of an un-ionized histidine portion (Figures 4.7 and 4.8). (Only the side chains of the two amino acids histidine and serine are shown located on the enzyme surface.) To assist in this way the histidine portion or residue must remain uncharged; it becomes ionized in acid conditions.

$$
\begin{array}{c}
H\quad H \\
C-N \\
N\diagup\quad\diagdown C \\
C \\
|\\
CH_2 \\
|\\
CH \\
H_2N\diagup\quad\diagdown COOH
\end{array}
\quad\underset{\xrightarrow{\;H^+\;}}{\rightleftharpoons}\quad
\begin{array}{c}
H\quad H \\
C-N \\
HN^+\diagup\quad\diagdown CH \\
C \\
|\\
CH_2 \\
|\\
CH \\
H_2N\diagup\quad\diagdown COOH
\end{array}
$$

ionization of histidine

At pH 5 it is so ionized that the reaction (Figure 4.8) does not take place, the acyl–enzyme intermediate is stable and reaction between enzyme and substrate is incomplete and stops at this stage.

Active sites

A molecule of chymotrypsin is made of a large number of amino acid units but only a small fraction of these play a direct part in the catalysis of substrate. A few are chemically important and many others provide a structural framework to provide ideal siting of the chemically active side chains and groups. All this may seem excessively complex for the mere breakdown of protein (something that can be done in the laboratory by simply boiling proteins with dilute acids). But the key to understanding the significance of enzymes lies in their substrate specificity. To break certain links in a chain without affecting the others requires an elaborate molecular machinery without which the orderly and complex chemistry of living organisms would be quite impossible. The action of chymotrypsin provides some insight into enzyme action at the molecular level.

Much of our knowledge about the molecular structure of chymotrypsin and other enzymes was obtained before X-ray crystallographic methods provided three-dimensional pictures of the molecules. Such representation of the structures is valuable; it does not tell us which parts of the protein molecule are actually associated with catalytic reaction but indicates that parts of the molecule are probably structural, somehow maintaining the active sites in their appropriate relative positions. Chymotrypsin contains 245 amino acid units and by

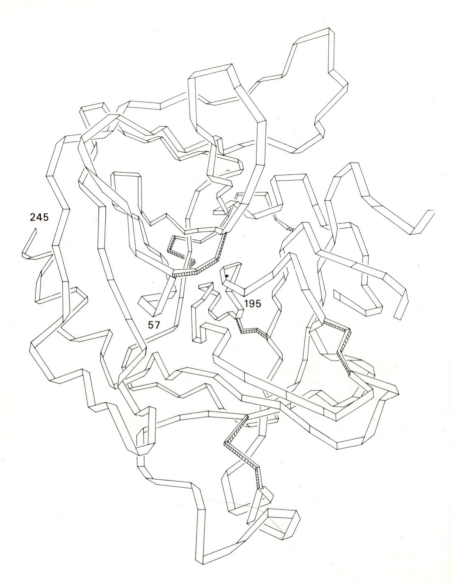

Figure 4.6. Diagrammatic representation of a molecule of α chymotrypsin

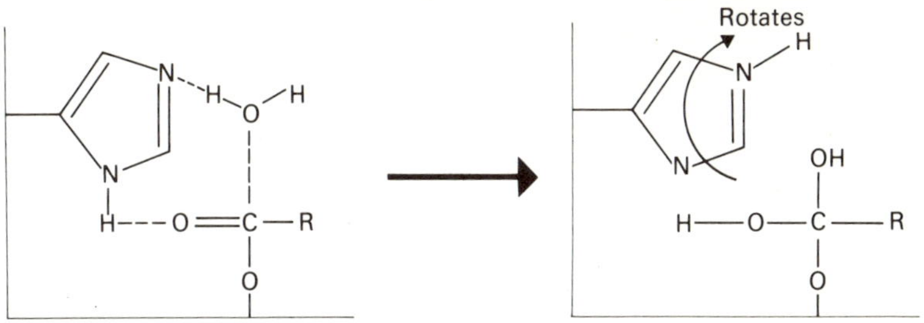

Figure 4.7. De-acylation of chymotrypsin

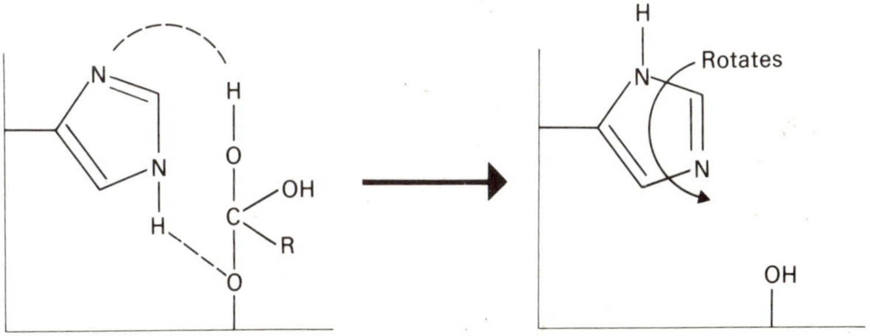

Figure 4.8. De-acylation of chymotrypsin

comparison with other enzymes seems to be a relatively small molecule. The three-dimensional structure is remarkably complex with many weak bonds between amino acid side chains (Figure 4.9).

The effect of pH

Weak interactions can be changed or modified by a variety of physical and chemical agents such as pH so that an enzyme is quite likely to possess an active conformation in only a limited pH range. In chymotrypsin this range is limited by the ionization of only two groups, the histidine at position 57 (see Figure 4.6), actually *chemically* involved in reaction and the terminal amino group. Active chymotrypsin requires an uncharged histidine and a charged terminal amino group. The terminal amino acid is not involved in reaction; its charge is necessary to preserve the active conformation of the enzyme. Thus, the affect of pH on enzyme activity does not necessarily indicate which amino acids are involved at the active site.

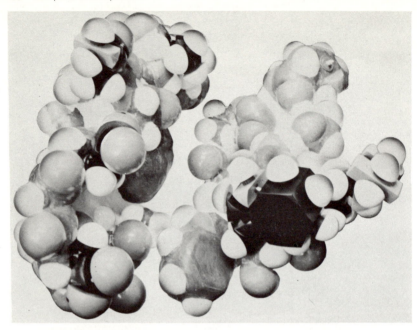

Figure 4.9. A model of a small part of a chymotrypsin molecule showing the probable spatial relationships of the atoms. The smallest white hemispheres represent hydrogen atoms, the larger grey ones, oxygen atoms, and the large, black hexagonal body represents the benzene ring part of the tryptophan (62) residue. Such a model gives some idea of how a substrate molecule with a complex shape might fit specifically into the active site of an enzyme

An amino acid group can be represented as RNH_2 and in the ionized form as RNH_3^+. Ionization takes place as follows

$$RNH_3^+ \rightleftharpoons RNH_2 + H^+$$

This equilibrium is governed, like other equilibria, by the relationship of concentrations then:

$$K_a[RNH_3^+] = [RNH_2] \cdot [H^+]$$

where K_a is the equilibrium constant, *or*

$$K_a = \frac{[RNH_2][H^+]}{[RNH_3^+]}$$

this constant is usually expressed (like pH) in powers of 10 as pK_a. For the terminal amino group of chymotrypsin pK_a for the equilibrium is about 8, that is K_a is 10^{-8}. If $[H^+]$ equals K_a then $[RNH_2]$ will equal $[RNH_3^+]$ and half of all terminal amino groups present will be ionized. This will occur at pH 8. At pH 7, 90 per cent of the amino groups will be ionized; at pH 8, 99 per cent and so on. Table 4.1 shows how much of

Table 4.1

pH	Charged amino (per cent)	Uncharged imidazole (per cent)
4	99.9	1
5	99.9	10
6	99	50
7	90	90 optimum
8	50	99
9	10	99.9
10	1	99.9

each of the essential forms of these two groups are present at particular pH values. From the table we can see at once that both essential forms are present optimally at pH 7 and this represents the pH optimum for this enzyme (as shown Figure 4.10). Because other ionizable groups in chymotrypsin are also modified by pH changes, the optimum is in practice rather sharper than is indicated by Figure 4.10. It is significant

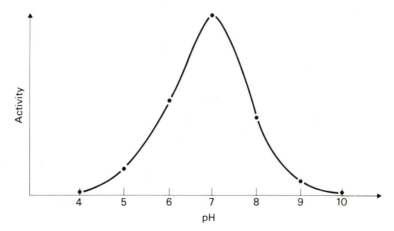

Figure 4.10. Activity/pH curve

that chemical modification of the terminal amino group leading to the much more basic amidine structure, provides a grouping which has a pK_a greater than 12.

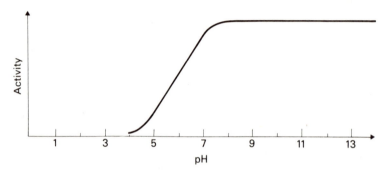

This modified chymotrypsin remains enzymically active right up to this pH. Beyond this value, chemical instability of both substrate and enzyme prevents measurements of activity (Figure 4.11).

Figure 4.11. Activity/pH curve

The effect of temperature

Enzymes are also susceptible to temperature changes. As we should expect, catalytic activity like other chemical activity tends to increase with a rise in temperature. For enzymes this is true only to a limited extent from about 20 to 40 °C (although some, like subtilisin, the washing powder enzyme, are known to be active outside this range). At higher and lower temperatures the structure of enzyme molecules is modified by alterations in the stability of the many weak interactions. Boiling enzyme solutions is an effective way of destroying catalytic activity completely. Thus with temperature as well as pH there is an optimum. For many enzymes it is about 37 °C the approximate body temperature of many animals including man. Temperature/activity curves are usually asymmetrical (Figure 4.12) as activity increases steadily with temperature to the optimum then falls away sharply with rapid inactivation above this point.

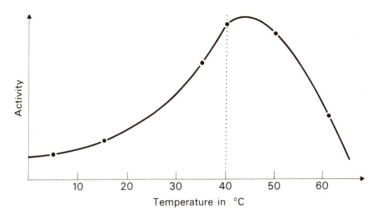

Figure 4.12. Activity/temperature curve

Poisons

High temperature inhibits enzyme action by modifying the molecular structure. Many chemical enzyme inhibitors are known and these have been a means of discovering more about enzyme structure. The more complex a system is, the greater the number of factors causing its failure or breakdown. This is as true for machines as it is for chemical reactions. Enzymes, viewed as molecular machines, are no exception. Besides responding to change in temperature and pH, enzymes are capable of chemical modification and this generally leads to a reduction in activity or inhibition. The actions of many poisons and drugs can be traced to their effects on specific enzymes. Poisonous compounds of metals such as arsenic, cadmium, lead, or mercury have a high affinity for thiol groups ($-SH$) in proteins, often part of the active site of an enzyme. Cyanide combines with respiratory enzymes containing copper and deactivates them.

Enzyme inhibitors

Inhibitors can affect the behaviour of enzymes in a variety of ways. Some combine with the enzymes and inactivate them completely and irreversibly. This effect has been applied to the battle against insects and as an instrument of war. Diisopropylfluorophosphate (DFP) as a constituent of pesticides and nerve gas completely inactivates chymotrypsin by attaching itself to the hydroxyl group of serine 195:

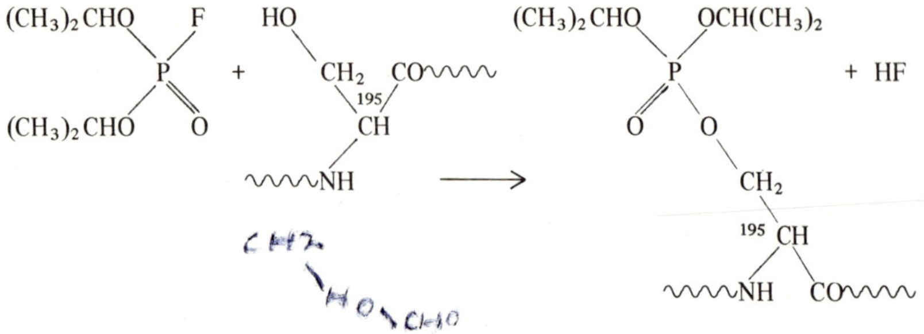

$$(CH_3)_2CHO \quad F \qquad HO$$
$$\diagdown \diagup$$
$$P \quad + \quad CH_2 \quad CO\!\sim\!\sim$$
$$\diagup \diagdown \qquad \backslash_{195}\diagup$$
$$(CH_3)_2CHO \quad O \qquad CH$$
$$\diagup$$
$$\sim\!\sim\!\sim NH$$

$$(CH_3)_2CHO \quad OCH(CH_3)_2$$
$$\diagdown \diagup$$
$$P \qquad\qquad + \; HF$$
$$\diagdown$$
$$O \quad O$$
$$\diagdown$$
$$CH_2$$
$$\backslash$$
$$^{195}CH$$
$$\diagup \diagdown$$
$$\sim\!\sim\!\sim NH \quad CO\!\sim\!\sim$$

$\longrightarrow$

[handwritten annotations: CH₂, HO—CH₃]

The phosphorylated serine residue is sufficiently stable to enable the chymotrypsin to be broken down with substituent still attached and this led to the tentative identification of this residue as an active site. The inhibition is stoichiometric, that is, it follows a simple rule of proportion; one molecule of DFP inactivates one molecule of chymotrypsin. The irreversible inhibition of chymotrypsin by DFP is very similar to that of an active serine residue in another enzyme, cholinesterase. Nerve impulses effectively pass across junctions by the production of a compound, acetylcholine, which is then broken down almost instantly by cholinesterase. Consequently if this enzyme is inhibited the nervous system of an animal is rendered useless. This is the basis for the use of DFP as an instrument of war; minute quantities absorbed into the body cause the body muscles to contract as if receiving continued nervous impulses.

In chymotrypsin only serine 195 reacts with DFP indicating that this serine is chemically different from or more active than the other 26 in the molecule. This emphasizes the need to remember the molecule as a whole which functions as chemically active sites held in a structure by chemically inactive amino acid residues.

Different types of inhibitor

Some inhibitors do not behave stoichiometrically as the effects do not seem clearly related to the amount present. Furthermore the action of inhibitors can be reversed by their removal by such physical means as dialysis. Attempts at classification of these reversible inhibitors, based on the effect on the Michaelis constant K_m or maximum velocity of reaction v_{max}, have not been altogether satisfactory. Terms like 'competitive', 'non-competitive', 'uncompetitive' inhibitor have been less than helpful.

If we consider enzyme catalysis to be essentially the result of two reactions

$$E + S \underset{k_2}{\overset{k_1}{\rightleftharpoons}} ES$$

and

$$ES \xrightarrow{k_3} \text{products}$$

where E = enzyme, S = substrate, and ES = enzyme–substrate complex, then inhibitors can be considered to belong to one of two categories:

(1) affecting the combination of E and S or
(2) affecting the rate of the overall reaction.

Category (1) inhibitors affect the substrate binding, and generally these inhibitors have a structural resemblance to the substrate; they compete with the substrate for the active site and are known, therefore, as 'competitive inhibitors'. For example, inhibitors of the enzyme monoamine oxidase are themselves amines. Competitive inhibitors of chymotrypsin include hydrocarbons such as naphthalene and D-amino acid derivatives of phenylalanine, tyrosine, and tryptophan (the optical isomers of the naturally occurring amino acids). Lineweaver-Burk plots (see p. 50) of enzyme activity are modified as shown in Figure 4.13 by competitive inhibitors. The value of K_m is increased as the enzyme is not able to react with substrate so easily. So $1/K_m$ is decreased by competitive inhibition; the maximum velocity is unchanged.

Category (2) inhibitors affect the catalytic step

$$ES \rightarrow \text{products}$$

and are known as non-competitive inhibitors. In theory they are supposed to affect k_3 and hence v_{max}.

Unless k_3 is very much smaller than k_2, then K_m is also affected by changes in k_3. Both maximum velocity and K_m are altered by non-competitive inhibitors as indicated in Figure 4.13(b) where both the theoretical and real situations are shown.

The greatest drawback to this system of inhibitor classification is the idea that, in a complex thing like an enzyme, it is possible to affect one part without altering others. Knowing their great structural complexity, such simple notions of inhibitor action are misleading. Enzymes are not only complex but highly integrated molecules, the product of 500

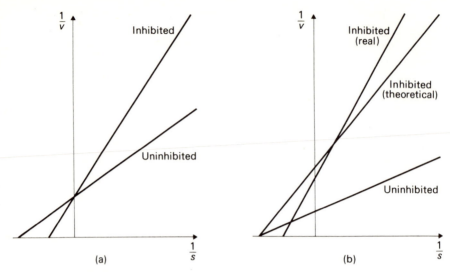

Figure 4.13. Lineweaver-Burk plots modified by competitive inhibitors

million years of organic evolution, in which *every* group of atoms plays its part.

Every discovery of new reactions involving chymotrypsin or any other enzyme causes speculation about its molecular structure. It is unreasonable to try and separate function from form and so the structure of enzymes is considered further in the next chapter.

5.

Enzyme Models

In recent years chemical and physical techniques (particularly X-ray crystallography) have provided much of the information necessary to understand the ways in which enzymes attach to and react with their substrates and to propose increasingly convincing pictures or models of enzymes to account for their chemical behaviour. The active site of an enzyme is visualized as a 'template' exactly fitting the substrate and providing just the right chemical attachments to enable it to react. Alternatively, the active site may be viewed as part of a catalytic surface (the enzyme) on which reaction takes place.

The lock and key model

These ideas were incorporated in the 'lock and key' model of enzyme specificity put forward by Fischer (1894) long before the chemical nature of enzymes was known. Enzymes can be seen as analogous to keys opening locks; the notched portion of the key corresponds to the active site. The key can open only one lock and thus the action is specific.

Such analogies of lock, key, and template reinforce the idea of an enzyme molecule, conveyed by X-ray crystallographic pictures, as a rigid structure. Consideration of three relatively simple enzymes may throw some doubt on this idea.

Example 1

Hexokinase catalyses the phosphorylation of glucose by ATP (Figure 4.2). The 6-OH of glucose is substituted by a phosphate group. The reactivity of the 6-OH group of glucose is very similar to that of the

OH group of water. Water molecules are much smaller than those of glucose and present in much higher concentration in the conditions in which the hexokinase reaction occurs. From these considerations alone, one might suppose that the omission of glucose would lead to the hydrolysis of ATP to ADP, the enzyme functioning as an ATP-ase instead of hexokinase in the absence of glucose, but this does not happen.

Example 2

For hexokinase, 2-O-methylglucose (see Figure 5.1) is neither an inhibitor nor a substrate, whilst 2-deoxyglucose is a substrate. This indicates that the large substituent in the 2-position does not permit 2-O-methyglucose to approach the enzyme. But N-acetylglucosamine is a *competitive inhibitor,* even though the acetamido (CH_3CONH-) group is much bigger than a single O-methyl (CH_3O-) group.

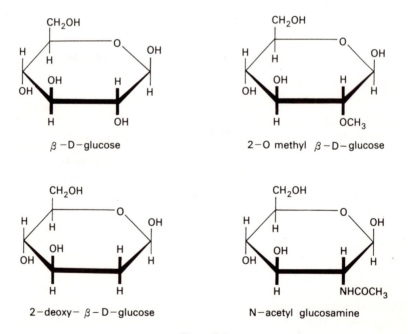

Figure 5.1

Example 3

Invertase (sucrose) catalyses the hydrolysis of sucrose. It is inhibited non-competitively by iodine so that v_{max} is greatly reduced but K_m not at all. According to the original theory a non-competitive inhibitor interferes with the kinetic step but not the binding of the substrate. But assuming the template model we are forced to ask how something as large as an iodine atom can affect the catalytic step without also upsetting the binding of substrate to enzyme.

Example 4

5'-nucleotidase has the power of splitting the terminal phosphate group from nucleotides such as AMP

adenosine monophosphate

For AMP (adenosine monophosphate) the K_m value is less than 10^{-4} and k_3 is very much less than K_2; v_{max} is 100. For nicotinamide ribose phosphate, K_m is 2×10^{-3} but v_{max} is virtually unchanged at 67. Nicotinamide is smaller than adenine and therefore less binding to the enzyme than might be expected. So for ribose phosphate, K_m should be greater still (having less affinity) and v_{max} should be unaffected. In fact, K_m is much the same (3×10^{-3}) as for the nicotinamide derivative, and v_{max} is reduced almost tenfold to 0.8.

Induced fit

These examples led Koshland (1959) to suggest the idea that enzymes are flexible structures and when the substrate combines with the enzyme it induces changes in shape or conformation so that the active groups of the enzyme are brought together. This is illustrated by Figure 5.2;

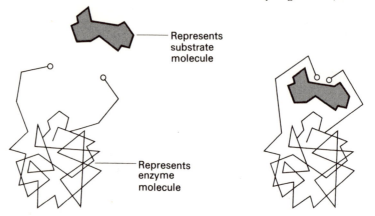

Figure 5.2. When the substrate combines with the enzyme it induces change of shape so that the active groups of the enzyme are brought together

Figure 5.3 shows why both larger and smaller compounds might be unsuitable for reacting with the enzyme. This idea was known as 'induced fit'.

Figure 5.3. Larger and smaller enzymes compounds are unsuitable for reacting with the enzyme

At about the same time as Koshland advanced his induced fit hypothesis, the structures of haemoglobin and myoglobin molecules were revealed through crystallographic and other studies. These substances are not enzymes but oxygen carriers in blood and muscle respectively. They are relevant here because the combination reaction with oxygen can be regarded as essentially similar to the first stage of an enzyme catalysis, the formation of enzyme–substrate. (We can regard them as 'honorary' enzymes.)

Substrate co-operativity

Oxygen binds to myoglobin through an iron atom in a complex ring system called haem (Figure 5.4). Oxygen binds to haem depending on the concentration of oxygen available. Uptake is thus a function of the partial pressure of oxygen and this can be represented graphically (Figure 5.5) as a simple kinetic curve, resembling Figure 4.1. Haemoglobin molecules contain four chains, structurally similar to myoglobin, each containing a haem group. These are two α chains of 141 amino acids

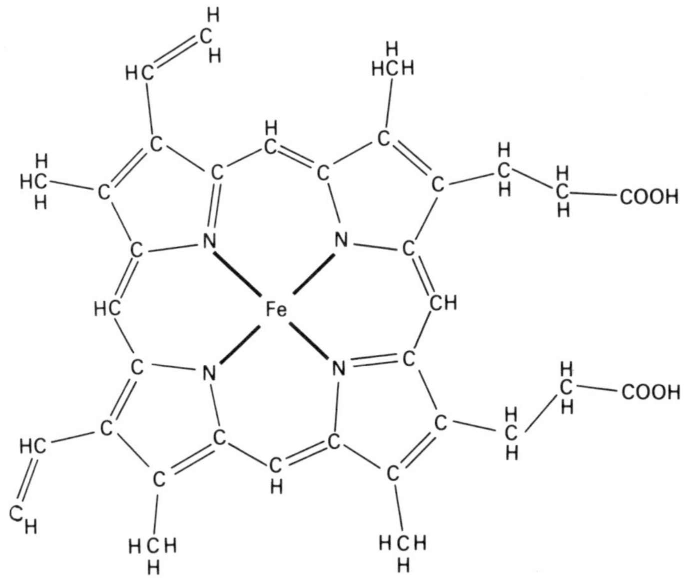

Figure 5.4. Haem

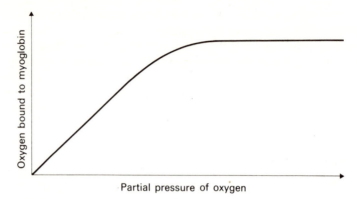

Figure 5.5. Uptake as a function of partial pressure for myoglobin

and two β chains of 146 amino acids each in the normal haemoglobin of healthy human beings. Haemoglobin has an oxygenation curve different from that of myoglobin, being S-shaped or sigmoid (Figure 5.6) so instead of the uptake of oxygen being directly proportional to the partial pressure (until saturation) as with myoglobin, there is a partial pressure range where more oxygen is taken up than would be expected from a simple linear relationship. This has been called *substrate co-operativity*; once one oxygen molecule is bound to a haemoglobin molecule it is easier for others to do so. Sigmoid relationships of this kind are known for a number of enzymes and substrates and this behaviour is known as *complex kinetics.*

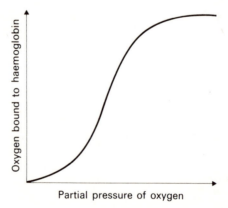

Figure 5.6. Uptake as a function of partial pressure for haemoglobin

Somehow there must be co-operative interactions between haem groups if we are to account for the sigmoid relationship. Oxygen binds to the haemoglobin at the iron atom of the haem group, but X-ray crystallography of haemoglobin reveals that the smallest iron–iron distance in a molecule is 2.5 nm. This distance is far greater than that over which direct chemical interactions are known to operate. There cannot be any *direct* haem–haem (iron–iron) interaction in haemo-globin! How, then, does one oxygenated haem group affect the oxygen of another?

Feedback inhibition

We can perhaps move towards an answer by examining other cases of cooperative binding of substrates. Threonine deaminase (TDA) catalyses the oxidative deamination of L-threonine to α-ketobutyrate and the rate/substrate concentration curve is sigmoid, indicating co-operativity.

$$\begin{array}{c} CH_3 \\ | \\ CHOH \\ | \\ NH_2-CH \\ | \\ COOH \end{array} \quad \xrightarrow{\text{TDA}} \quad \begin{array}{c} CH_3 \\ | \\ CH_2 \\ | \\ CO \\ | \\ COOH \end{array}$$

The reaction is the first of five stages leading to the biosynthesis of the amino acid L-isoleucine:

$$\begin{array}{cccc} CH_3\;CH_3 & CH_3\;CH_3 & \begin{array}{c}CH_3\\|\\CH_2\\|\\CH_2\\|\\CH_2\end{array} & CH_3\;CH_2 \\ \diagdown\diagup & \diagdown\diagup & & \diagdown\diagup \\ CH & CH & & CH \\ | & | & & | \\ NH_2-CH-COOH & CH_2 & NH_2-CH-COOH & NH_2-CH-COOH \\ & | & & \\ & NH_2-CH-COOH & & \\ \text{valine} & \text{leucine} & \text{norleucine} & \text{isoleucine} \end{array}$$

Curious as it may seem, isoleucine is an inhibitor of threonine deaminase and this is an example of 'feedback' or 'end-product' inhibition, where the final product of a metabolic pathway inhibits the first reaction. This inhibition is suppressed by mercury ions as are the sigmoid kinetics, though threonine deaminase remains active. In the presence of these ions an optical isomer of L-threonine, allothreonine, remains a theoretical, competitive inhibitor. Norleucine, though not an inhibitor of TDA, removes the isoleucine inhibition from it.

These observations provide evidence that:

(1) TDA shows co-operative binding of the substrate, like haemoglobin and oxygen, and
(2) the activity of TDA can be modified by a substance, isoleucine, which binds at a separate site to that of the substrate.

The question posed in Example 3 on invertase can be turned around—how can an inhibitor, binding at a site different from the active site, exert any inhibitory influence at all?

Oligomers

A feature of enzymes showing sigmoidal kinetics is the presence of more than one active site in each molecule. This can occur when an enzyme molecule consists of a number of similar sub-units or *monomers* each of which has an active site. Aggregations of sub-units have been named *oligomers*. In some cases there may be additional sub-units for the binding of substances which act as inhibitors or stimulators (effectors) of enzyme activity. These aggregations or oligomers may be broken into their constituent monomers by gentle heating or by the presence of mercury salts and other agents, and a consequence of this is the disappearance of co-operativity. Myoglobin can be regarded as a sub-unit of haemoglobin and it lacks the complexity in oxygen uptake shown by the larger aggregation (Figures 5.5 and 5.6).

There is now considerable evidence to support the hypothesis that co-operativity and effector action are brought about by changes in shape of monomer units. These changes are called allosteric transitions. The proteins in which these are said to take place are allosteric proteins and effectors of the kind described above, allosteric effectors or allosteric *ligands*.

Details of changes in molecular shape are often difficult to define from the results of complex analytical techniques. In haemoglobin molecules, X-ray evidence indicates that oxygenation of haemoglobin is associated with the two β chains moving 0.7 nm closer together. Exactly why such movement should lead to co-operativity is still obscure. This kind of movement is not restricted to substances showing co-operativity. Carboxypeptidase does not do so, yet X-ray studies show that parts of the protein move as much as 0.8 nm when the substrate binds to the enzyme; certainly this is evidence in favour of the simple induced fit hypothesis. There is little doubt that allosteric effects play an important part in the control of metabolism. Allosteric proteins may be a means of control through allosteric effectors, co-operativity, and the type of sub-units involved.

Control by inhibition

One aspect of metabolic control is the prevention of an accumulation of wasteful metabolites. There are a number of well documented cases of feedback inhibition which exert this kind of control. A potentially wasteful process is exemplified by glycogen metabolism. The synthesis and breakdown of glycogen (Figure 5.7) occur by separate pathways and if all the enzymes involved are active simultaneously in a cell the net reaction is merely the hydrolysis of the glucose activator uridine triphosphate. The three reactions are:

glucose-1-phosphate $+ UTP \rightarrow UDP -$ glucose $+$ pyrophosphate

$UDP -$ glucose $+$ glycogen$_n \rightarrow UDP +$ glycogen$_{(n+1)}$

glycogen$_{(n+1)} +$ phosphate $\rightarrow$ glycogen$_n +$ glucose-1-phosphate

The net reaction is: UTP to UDP + phosphate. However, the enzymes,

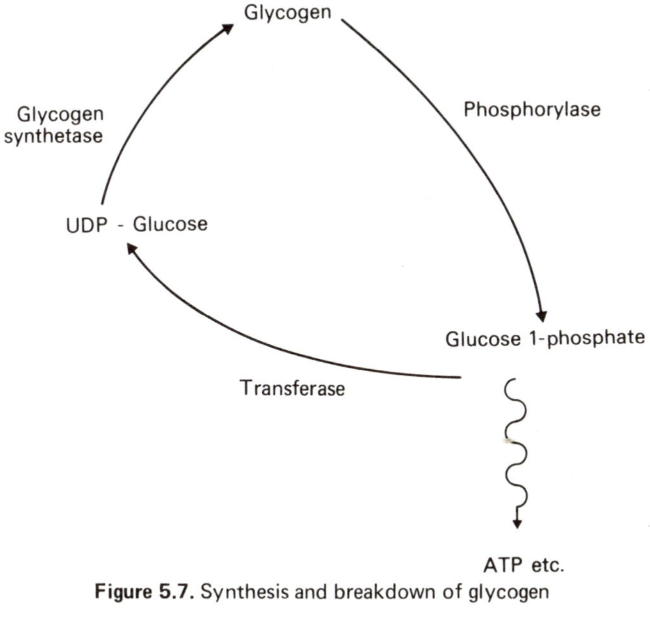

Figure 5.7. Synthesis and breakdown of glycogen

glycogen synthetase and phosphorylase, involved in these reactions, are both allosteric proteins reacting in opposite ways to allosteric effectors such as ATP, AMP, and glucose-6-phosphate. The two enzymes are also capable of chemical modification by enzymic phosphorylation and dephosphorylation. This too affects the activity and the state of aggregation of the sub-units of the enzymes and, together with other more complex interrelationships, results in the prevention of simultaneous reaction and the wasteful and fruitless hydrolysis of uridine triphosphate.

Metabolic stability

A necessary feature of a successful, complex set of interactions, such as that found in living cells, is stability. If each chemical constituent fluctuated rapidly in concentration from one moment to the next the system of metabolic pathways would become chaotic. The relationship between substrate concentration and rate of reaction for an allosteric enzyme is shown by Figure 5.8. Suppose that in the 'normal' condition of metabolism the substrate concentration is A (under the curved part of the line) and for some reason the concentration increases to A''. This would cause a disproportionate increase in the rate (v'') resulting in the rapid consumption of substrate and reducing it to the normal value A. On the other hand, if there is a decrease in substrate concentration (to A') there will be a fall in rate (v') so that *turnover* of substrate is still maintained at a nearly normal rate. The stabilizing effect of this

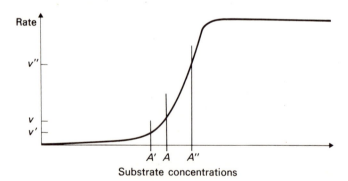

Figure 5.8. Rate of reaction and substrate concentration for an allosteric enzyme

arrangement hinges on the value of normal substrate concentration being located at A at the curved region of the line.

Some allosteric proteins are composed of one kind of sub-unit only; some, like haemoglobin, have two kinds, and others have a composition dependent on the *availability* of sub-units so that one enzyme may have multiple molecular forms. An example of an enzyme having multiple molecular forms is lactate dehydrogenase (LDH); it catalyses the reaction pyruvate to lactate

$$
\begin{array}{ccc}
CH_3 & & CH_3 \\
| & & | \\
CO & \rightleftharpoons & CH(OH) \\
| & & | \\
COO^- & & COO^- \\
\text{pyruvate} & & \text{lactate}
\end{array}
$$

which is a metabolic shunt because further reaction of lactate has to be preceded by reconversion to pyruvate. Pyruvate could be described as being near the crossroads of cell metabolism because it is an intermediate in the chemistry of fats, carbohydrates, and amino acids (Figure 5.9). Its further metabolism via acetyl-CoA leads to the release of much energy via the respiratory chain of reactions (see Chapter 3). It is a process

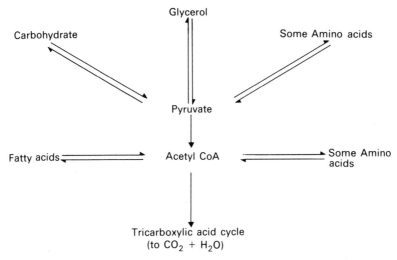

Figure 5.9. Pyruvate can be considered to lie at the crossroads of fat, carbohydrate, and amino acid metabolism

which requires oxygen for its completion. In highly oxygenated tissue, such as heart muscle, the formation of lactate does not take place and much more energy is released in the further metabolism of pyruvate than, for example, the initial metabolism of glucose. Lactate dehydrogenase of heart muscle is much less active than that of less oxygenated tissues like liver or skeletal muscle. Both these tissues produce considerable quantities of lactate; the accumulation of this in muscles is supposed to be the cause of cramp associated with muscular fatigue.

Isozymes

Lactate dehydrogenase is composed of four sub-units of two kinds, A and B. Liver LDH is composed almost entirely of type A sub-units, whereas in heart tissue LDH of type B sub-units predominates. The two forms of enzyme are termed LDH_5 and LDH_1 respectively. Other forms (see Table 5.1) are distributed in tissues according to the energy requirements and oxygen supply.

Table 5.1

LDH isoenzymes (or isozymes)	No. of sub-units	
	Type A	Type B
LDH_1	0	4
LDH_2	1	3
LDH_3	2	2
LDH_4	3	1
LDH_5	4	0

The isozymes can be distinguished by relatively simple laboratory techniques such as electrophoresis, and this has led to the use of this enzyme in diagnosis. Damaged tissues leak intracellular materials including LDH into the blood stream. The type of LDH in the blood is indicative of the organ from which it has come and knowledge of which organ is damaged (and this is often difficult to determine by other means) can be a great aid in diagnosis and treatment of a patient.

The availability of the sub-units which govern the assembly of isozymes is thus related to the type of tissue under consideration and this, in turn, is genetically determined. Deoxyribose nucleic acid (DNA) in cell nuclei is a controlling agent of tissue type and occasionally the control process goes wrong. For example, blood formation is controlled by the nuclei of cells in the bone marrow which divide and differentiate to produce red cells or erythrocytes in the blood stream. Occasionally a mistake in the DNA-controlling process causes a lack of α chains and consequently the formation of peculiar haemoglobin composed entirely of β chains. This unstable haemoglobin leads to the formation of modified erythrocytes—so called haemoglobin H disease. These erythrocytes are removed from the blood stream more rapidly than the normal type and severe anaemia results.

Though the availability of protein is ultimately under the control of DNA within the cell nuclei, the examples quoted above are enough to indicate that a considerable part of the metabolic management in a cell is controlled by the proteins themselves.

6.

Oxidation and Reduction

Although vast numbers of reactions go on in even the simplest organisms, there are not so many different *types* of reaction. The most important and significant are those involving the transfer of hydrogen atoms (oxidation/reduction) and phosphate groups (esterification/hydrolysis) from one molecule to another.

Unravelling many complex systems of metabolism has revealed many sophisticated molecules which act merely as donors and acceptors of hydrogen or of phosphate. Respiration, for example, can be regarded, in part, as the transfer of hydrogen from one acceptor to another; each reaction being catalysed by its own specific enzyme. The ultimate fate of such hydrogen is combination with oxygen to form water. It can be said that we breathe air (oxygen) merely to remove hydrogen from our metabolic system, as water.

Energy transfer

It is not always easy to see the point of such hydrogen transfer. In isolation it appears only as a means of degrading food materials by oxidation with very little output of heat. If it serves as a means of energy transfer, the process is by no means obvious at first glance.

The removal of hydrogen from any substance whatever is termed oxidation. The departure of one or more hydrogen atoms from the molecule of a compound means that electrons are removed from shared orbits. Thus, removal of electrons also constitutes oxidation. It is by oxidation of another kind that energy is usually obtained from fuels. Oxygen combines directly during combustion and the heat emitted is trapped by some physical means and made to expand gas or generate steam so that useful work can be done. Glucose like other food substances is oxidized by organisms and this can be simply expressed by the reaction:

$$C_6 H_{12} O_6 + 6O_2 \rightarrow 6CO_2 + 6H_2 O$$

but this single reaction certainly does not take place in metabolism.

Electrical energy

The concept of oxidation as something other than the direct combination with oxygen is vital to our understanding of metabolism in general and respiration in particular. The connection between oxidation and energy transfer can be usefully explored in another direction by reference to simple electrical cells.

It has been stated that oxidation is the removal of hydrogen atoms from a molecule or, simply, the removal of electrons from an atom. Suppose a piece of zinc is placed in a solution of copper sulphate, the zinc begins to dissolve and this can be expressed as:

$$Zn + CuSO_4 \rightarrow ZnSO_4 + Cu$$

$$Zn + Cu^{++} \rightarrow Zn^{++} + Cu$$

The zinc is thus oxidized and the copper reduced because electrons have departed from the zinc producing positively charged ions (the reaction has a free energy of -214 kJ mol^{-1} and is favourable). Where ϵ represents an electron:

$$Zn \rightarrow Zn^{++} + 2\epsilon$$

$$Cu^{++} + 2\epsilon \rightarrow Cu$$

If these two half-reactions are spatially separated but connected by electrical conductors then a simple electrical cell is formed.

By using copper, zinc, and solutions of their sulphates we can make a Daniell cell, see Figure 6.1, which produces an electromotive force (EMF) of approximately 1.1 V. The porous pot separates the two half-reactions but provides a conducting path for electrons (electric current). The terminals can be joined to an electric motor and useful work performed. The work is derived from an oxidation reaction but without involving oxygen, or the production or utilization of heat.

Very many other half-reactions can be set up; the Daniell cell is merely one application of such reactions, of passing interest. In order to compare all the feasible half-reactions a standard hydrogen electrode can

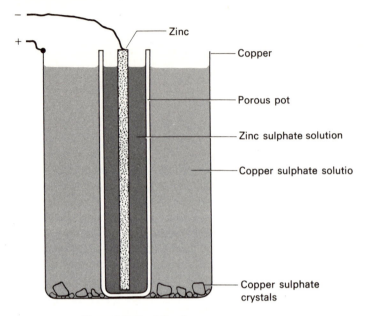

Figure 6.1. Daniell cell

be used as a reference. This is made by bubbling hydrogen over a platinum electrode immersed in a 1 M-solution of hydrogen ions (1 M-HCl) (Figure 6.2)

$$\tfrac{1}{2}H_2 \rightarrow H^+ + \epsilon$$

and the *electrode potential* is defined as zero. Every other half-reaction has a fixed positive or negative value compared with that of the hydrogen electrode. For example the zinc–(M)zinc sulphate electrode coupled to the hydrogen electrode has an electrode potential of −0.76 V; for copper–copper sulphate it is +0.34 V.

In a Daniell cell the metals not only take part in reactions, they also act as conductors of electrons. Many oxidation–reduction reactions do not involve metals in this way; for example, the ferricyanide–ferrocyanide system and the cytochrome b_5 system

$$Fe(CN)_6^{3-} + \epsilon \rightarrow Fe(CN)_6^{4-}$$

$$Cyt\ b_{5ox}^{3+} + \epsilon \rightarrow Cyt\ b_{5red}^{2+}$$

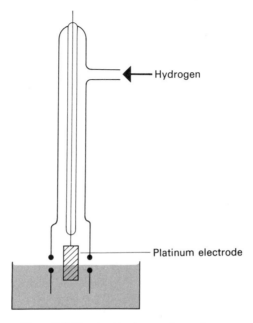

Figure 6.2. Standard hydrogen electrode

Appropriate electrodes can be made by inserting inert conductors such as platinum into mixtures of the solutions. In such systems, the concentration at the *standard* electrode potential (E°) is defined when the ratio of the participating species is one. For cytochrome b_5 this is +0.02 V and for the ferricyanide–ferrocyanide system it is +0.36 V. As the ratio of oxidized to reduced cytochrome b_5 becomes *smaller,* so also does the redox potential. Starting with virtually complete reduction (all the cytochrome b_5 reduced), the E value is seen to rise through +0.02 V to complete oxidation in Figure 6.3. A change from 99 per cent reduction to 99 per cent oxidation is associated with a redox potential change of just over 0.2 V.

Something is needed to oxidize the cytochrome; ferricyanide is very suitable (low activation energy) and higher redox potential. In this case the overall reaction is:

$$\text{Cyt } b_5^{2+} + \text{Fe(CN)}_6^{3-} \rightarrow \text{Cyt } b_5^{3+} + \text{Fe(CN)}_6^{4-}$$

If we start with reduced cytochrome and add ferricyanide, the value of the potential difference between the two redox potentials ΔE, since two

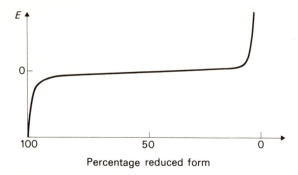

Figure 6.3. The change from 99 per cent reduction to 99 per cent oxidation is associated with a redox potential change of 0.02 volts

E values are involved, changes as indicated in Figure 6.4. The E° value for the ferricyanide–ferrocyanide system is +0.36 V and at the equivalence point (where reaction is complete) the ΔE value is halfway between the two E° values, +0.17 V. If a small quantity of a redox *indicator* with an E° of about this value is added it will change colour at the equivalence point; cresyl violet, for example, with E° = +0.17 V can be used to determine the end-point in the titration of reduced cytochrome b_5 with ferricyanide. Alternatively, the actual colour of the cytochrome can be used, since the reduced form absorbs light of

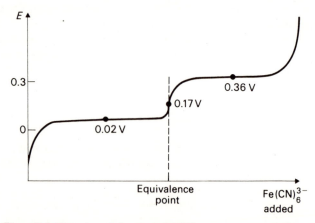

Figure 6.4. The value of the potential difference between the redox potentials when ferricyanide is added to cytochrome

wavelengths 423, 526, and 556 nm, whilst the oxidized form only absorbs at 413 nm. The change from red to yellow can be measured in a spectrophotometer and this is a common way of investigating such reactions in biological systems.

In passing, we may note that the chemical basis of the oxidation–reduction system of cytochrome b_5 is iron; the change from oxidized to reduced is associated with an Fe^{+++} to Fe^{++} conversion. Iron or copper form the chemical bases of all the cytochromes, whose E° values range from 0.02 to 0.29 V; in free solution the E° value of the Fe^{+++} to Fe^{++} system is 0.77 V, which indicates how the binding of a metal to protein changes its chemical properties.

The electrode potentials or *redox potentials E* provide information of a theoretical kind about which substances will oxidize and which reduce (Figure 6.5). The *differences* between redox potentials (ΔE), like free

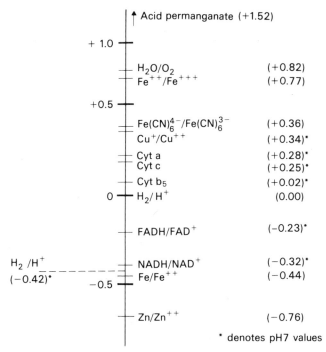

Figure 6.5. Redox potentials at 25 °C. Note: a system standing more than 0.2 V above another should be able to oxidize the reduced form

energy values, tell us if reactions are feasible, not that they actually take place. It may be, for example, that the activation energy is too great to allow some feasible reaction to occur. Such reactions may be brought about if the appropriate catalyst or enzyme is present. There is a direct quantitative relationship between ΔE and ΔF. Just as the free energy changes in reactions depend upon the concentrations of both reactants and products, so electrode potentials depend upon the concentrations of the participants in an oxidation–reduction system. An electrical cell does not last for ever, since the chemical constituents are used up; the Daniell cell has to be 'fed' from time to time with copper sulphate to maintain its EMF.

The invention of the Daniell electrical cell was part of a search for energy. A modified form of a slightly different electrical cell (the Leclanché cell) is widely used today as the 'dry battery' for flashlights and transistor radios. We are concerned to understand how living cells derive energy from food materials, energy which can subsequently be used to drive chemical reactions within the living cell.

Cytochromes and the Krebs cycle

Cytochromes and other substances which undergo similar oxidation–reduction play a vital part in cell respiration. The first part of this long and involved sequence of changes is glycolysis, which is responsible for the breakdown of glucose and other food materials to pyruvate and the formation of some ATP. The pyruvate, through acetyl CoA, marks the beginning of a further complex series of changes, the tricarboxylic acid cycle (or Krebs' cycle). This cycle (Figure 6.6) though apparently important seems to involve little energy transfer and its true significance would appear to be the linking of glycolysis to the respiratory chain of cytochromes and other substances, which not only results in the final products of food oxidation (carbon dioxide and water) but also the transfer of energy. The tricarboxylic acid cycle has significance as a means to an end, not an end in itself.

In animal tissues it has been found that the tricarboxylic acid cycle leads to a reduced form of nicotinamide adenine dinucleotide (NADH) which is reoxidized to NAD^+ via a chain of cytochromes in the mitochondria. Experimental measurements of redox potentials have revealed a ΔE value of 1.14 V for the oxidation corresponding to a free energy change of 218 kJ mol^{-1}. This energy is released in smaller 'packets'. (NAD is nicotinamide adenine dinucleotide.) The energy gaps,

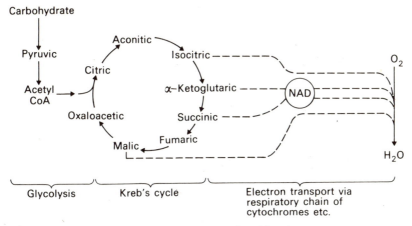

Figure 6.6. Krebs tricarboxylic acid cycle

in some cases, are enough to accommodate the synthesis of an ATP molecule from ADP and phosphate. Three such gaps or places have been suggested to account for the fact that three ATP molecules are synthesized for each NADH molecule oxidized. The molar ratio between phosphate used and oxygen consumed seems well established (P/O is 3), but no intermediates, involving cytochrome systems of linked oxidation and phosphorylation, have yet been isolated after 20 years of concentrated endeavour. It is possible that the idea of *chemical* intermediates is a false hypothesis and alternatives have been suggested.

The chemiosmotic hypothesis

One alternative uses concepts associated with the electrical cell. It was pointed out that in a successful cell there must be spatial separation of two half-reactions. Mitchell in 1966 put forward a chemiosmotic hypothesis which envisages a spatial separation of various half-reactions of the cytochrome chain by mitochondrial membranes. Many biological half-reactions involve hydrogen ions (a hydrogen atom without its single electron consists of a proton) and one such is the final formation of water by the reduction of oxygen:

$$\tfrac{1}{2}O_2 + 2H^+ + 2\epsilon \rightarrow H_2O$$

If such a reaction occurs on one side of a membrane which is impermeable to protons, the solution round about would become deficient in protons. If the membrane contains an electron carrier (M), then one reaction could be linked to another reaction without the need for an intermediate of the kind which has, so far, defied isolation and identification. The idea of a membrane linkage can be represented diagrammatically (Figure 6.7). A membrane impermeable to protons would thus generate a proton gradient across it, a proton-motive force, and provide protons for ATP synthesis:

$$ADP^{3-} + HPO_4^{2-} + H^+ \rightarrow ATP^{4-} + H_2O$$

The synthesis of ATP would occur via another membrane channel, recycling the protons generated by the electron carriers (cytochromes) (Figure 6.7). It has been calculated that a pH difference (proton gradient) of 3.5 units across the membrane would be enough to drive this reaction in normal, physiological conditions, and some experimental evidence for such a 'proton gradient' across the mitochondrial membrane exists. Similar mechanisms are believed to apply to ATP synthesis in the photosynthesizing chloroplast but in these the situation is less clear.

All the oxidation–reduction reactions outlined above require specific

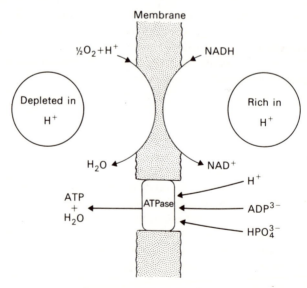

Figure 6.7. Membrane linkage

enzymes to catalyse them as do all the respiratory reactions so far discovered. In the cytochrome chain the cytochromes themselves are an intimate part of the enzyme systems. Oxidation–reduction enzymes are usually associated with a co-factor which undergoes oxidation and reduction during reaction. Most of these co-factors contain ions of transition metals. For example, some cytochromes contain iron in the same porphyrin system as is found in haemoglobin.

The organization of oxidation

Organisms remain alive so long as their metabolism continues as a multitude of orderly and interdependent reactions. Even the so-called physical actions such as movement are due to muscle contraction which is essentially a chemical rearrangement of actin, myosin, and other substances of which muscle is composed. We live because the 'right' reactions occur in the right place at the right time. Many of these would never occur (even in the presence of an appropriate enzyme) outside the organism because in isolation they are thermodynamically unfavourable. In the test tube we have to apply energy to encourage the reaction, but in nature success is achieved by coupling unfavourable reactions to those which have an appropriate free energy change. Enzymes play an essential part in this for it is believed that they are arranged in an orderly fashion in structures such as mitochondria and possibly separated by membranes in order to preserve the sequence necessary for successful, serial reaction.

The sequence known as aerobic respiration is the stepwise oxidation of food materials. In this oxidation we see no single energy-releasing combination of food and oxygen. The model of a man-made heat engine will not do. Electrical cells remind us that useful work can be done by oxidations with little or no release of heat and no direct participation of oxygen. It may be that some features of living cells are similar to those in the electrical cell but the analogy does not take us far. Oxidation by respiratory metabolism results not in the evolution of large quantities of heat or electrical energy but in the production of useful chemical compounds, particularly ATP. This coupling of oxidation–reduction reactions to phosphate transfer (essentially hydrolytic reactions involving water elimination) has no parallel in the test-tube and clearly represents a high degree of chemical sophistication. The total process is equivalent, energetically, to direct combination of food and oxygen but the subtlety and multiplicity of the organic processes ensure that the available energy is exploited in a way which can be contained within the delicate environment of living cells.

7.

The Significance of Enzymes

Because enzymes function as catalysts, there is a tendency for them to be regarded as merely agents that facilitate reaction, that accelerate chemical change without actually altering the products or the equilibrium between them and the products. Though this is true, the significance of enzymes in metabolism is gravely diminished unless certain additional facts are borne in mind. Reactions occur in organisms when the reactants are available, when the energetics of the system are favourable, and when the appropriate enzymes are present. The occurrence of an enzyme at a particular time and place within a cell is enough to ensure that its associated reaction takes place. In a sense, such an enzyme can be said to cause the reaction. This is a much stronger and more significant notion than mere facilitation of a reaction which would take place anyway.

Enzymes as organizers

It should be noted also, that enzymes are not consumed or decomposed in reaction and that they may be fixed or tethered to a cell organelle such as endoplasmic reticulum or a mitochondrion. Orderly arrangements and sequences of enzymes make possible complex, multistage chemical processes resulting in compounds quite beyond the scope of a 'brew' of reactants and catalysts mixed randomly in a liquid medium. The term 'compound' may perhaps suggest metabolites remote from common experience such as ATP and DNA, but it must not be forgotten that wood, skin, blood, and bone are all chemical 'compounds' and all the products of chemical reactions. They have been made because enzymes, somewhere and at some time, have been at work. Put more generally, an organism, structurally as well as functionally, is the product of enzyme action.

The origin of enzymes

This idea leads to an obvious question. If organisms are the products of enzyme action, what makes the enzymes in the first place? This is one aspect of the more general and fundamental question 'how does an organism make itself—by what means does it develop?'

There is a direct link between enzyme action and the carrier of genetic information, referred to previously, deoxyribose nucleic acid (DNA). Much has been written elsewhere about the way DNA carries this information. There is now convincing evidence to support the idea that the incredibly long chain molecules of DNA arranged as a double helix form a code composed of sequences of four chemical units, thymine, adenine, cytosine, and guanine. These units are arranged linearly in such a way that groups of three such as GCG or AGC etc. (called codons) can act as code words or call signs corresponding to individual amino acids. The message made of a sequence of codons can be replicated and this is done before each cell division. Furthermore, shorter lengths of the code can be copied on to ribose nucleic acid (RNA) molecules and translocated from the nucleus where DNA resides to the cytoplasm where protein synthesis takes place.

The important step in the present context is the linear arrangement of amino acids called into place by the coded message from DNA. Such a sequence can be condensed into a protein molecule. Proteins differ from one another because they are made of different selections of the twenty or so amino acids and also because of the order of amino acid residues in the molecule. This order determines the final shape of the whole molecule and thus its chemical properties. Enzymes are proteins. Chymotrypsin, for example, is chymotrypsin rather than some other protein because of the sequence of amino acid residues in the molecule. Replacement and substitution of some of the residues might not alter the properties of the enzyme but alteration of those associated with the active site would render chymotrypsin useless as a digestive enzyme.

Enzymes hold a commanding position in the chemical arrangements within an organism. Cells make and maintain themselves from a supply of raw materials and by reference to a 'master plan' held by the DNA. But as every chemical change is controlled by an enzyme, these essential tools of metabolism form the link between the plan and its realization (Figure 7.1).

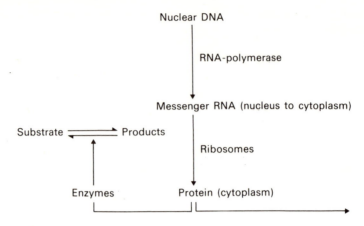

Figure 7.1. Enzymes form the link between plan and realization

Mutation

The sequence of nucleotides in DNA can be regarded as fixed and self-reproducing, but every now and then 'accidents' occur which result in a change of sequence—such changes are called mutations. Certain chemicals and atomic and ultra violet radiation (called mutagens) cause these 'accidental' alterations and modifications. (The bases thymine and cytosine seem particularly susceptible to alteration.) By no means all mutations lead to changes of amino acids or their sequence in proteins, and even when they do, the change may not cause serious alteration of the protein. Some mutations found in the β chain of human haemoglobin are shown in Figure 7.2. The most prominent changes (histidine to tyrosine and valine to glutamine) are both changes in the kind of amino acid; a basic hydrophilic to a neutral hydrophobic one and, in the second case, from a hydrophobic to a hydrophilic acid form. The other changes outlined have little, if any, deleterious consequences and are changes of like to like (except perhaps for J. Cambridge). The genetic code seems to have evolved so as to minimize the effects of mutagens and preserve stability in metabolism. Three of the amino acids mentioned (valine, glycine, leucine) are associated with triplet codes in which changes of the last nucleotide have no effect whatever.

Because mutations are changes in DNA they are inherited. A mutation occurring in a body cell is likely to affect only that cell and daughter cells formed by subsequent division. A mutation within a germ cell

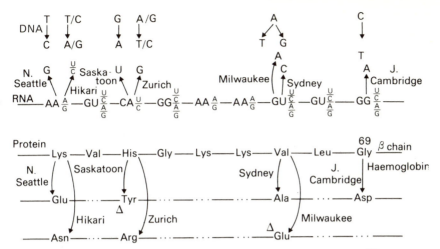

Figure 7.2. Some mutations of the β chain of human haemoglobin. (The mutations are usually named after the hospital in which they are first noticed)

Δ = serious pathological consequences.
Bases

A	adenosine monophosphate
T	thymidine monophosphate (in DNA only)
C	cytidine monophosphate
G	guanosine monophosphate
U	uridine monophosphate (in RNA only)

Amino acids

Lys	lysine
Val	valine
His	histidine
Gly	glycine
Leu	leucine
Glu	glutamic acid
Arg	arginine
Ala	alanine
Asp	aspartic acid
Asn	asparagine

which takes part in reproduction will be passed to every cell of the resulting offspring. Even so, effects of the mutation may not be seen—the condition is said to be recessive. (For an account of genetics and biochemistry see Harris, H., *The Principles of Human Biochemical Genetics*, North Holland Publishing Co., 1970.)

The conversion of phenylalanine to tyrosine has been mentioned earlier (p. 28). This reaction is catalysed by the enzyme *phenylalanine hydroxylase*. In the inherited disease phenylketonuria we believe that this enzyme is made inactive by an amino acid substitution—the result of mutation in the DNA. Accumulation of phenylalanine in the blood-stream occurs, leading to irreversible brain damage after only a few weeks of life. A diet almost completely free from proteins containing phenylalanine prevents this from happening. Such brain damage, in the past, led to phenylketonurics being a significant number of the institutionalized insane. Phenylketonuria is an example of an *inborn error of metabolism*; overall, roughly 0.1 per cent of births show indications of this kind of error, of which some 400 different types have been documented. It is believed that nearly all of these are the result of amino acid substitutions in proteins but definitive evidence on this point is lacking since the enzymes concerned are neither as accessible as haemoglobin in blood nor have simple enough methods been discovered for their purification and analysis.

Evolution

If a mutation produces a detectable effect, such as the malfunction of an enzyme, in an organism it is likely to be deleterious. This is not surprising; we would not expect a clock, for example, to be improved by a single, chance alteration to one of its working parts. But we believe that on rare occasions mutations result directly or indirectly in the improvement of a species. Modifications of DNA are the ultimate source of that inherited variability noted by Darwin. In his *Origin of Species* he remarked '. . . if variations useful to any organic being do occur, assuredly individuals thus characterized will have the best chance of being preserved in the struggle for life; and from the strong principle of inheritance they will tend to produce offspring similarly character-ized . . .'. That we now know that variability is inherited and achieved through proteins in no way diminishes the force of natural selection.

The basis of evolution is the modification of DNA leading to changes in the amino acid sequences in proteins. If the haemoglobins of sheep and cows are compared it is found that the amino acid sequences differ slightly. Of the 140 or so amino acids in each chain only eight have a common, consistent sequence in all species. Presumably mutations causing modification of these eight 'essential' portions are heavily selected against since, presumably, the haemoglobin does not function without the eight essential sites. But outside these eight, considerable variation is possible. Through the aeons of evolutionary time we can assume that great divergences of protein and enzyme structure have taken place which may have led to marked differences in enzyme activity.

So much attention has been given to evolution of organic structure; we must not forget that metabolic pathways must have evolved too. A small example of divergence is provided by the parasitic worm *Ascaris lumbricoides* which possesses an enzyme called phosphoenol pyruvate carboxykinase which is important in the biosynthesis of amino acids such as alanine and this operates at pH 6–7. The same enzyme in its host, man, operates at pH 7–8. In addition it has been found that the human enzyme is dependent on the presence of magnesium but the ascarid enzyme is zinc-dependent. Such differences provide possible means of specific chemical attack on the parasite without harming the host. Knowledge of these subtle evolutionary differences is extremely valuable because parasites well adjusted to their hosts are notoriously difficult to treat, and they afflict, in one form or another, two-thirds of the world's population causing debilitation though seldom killing their hosts outright.

Enzymes in a wider context

Attempts to understand metabolism and the part played by enzymes may prove to be more than an academic exercise. A number of useful applications of enzymology are already in daily use in laboratories and in industry but these have not yet made a major contribution to modern technology. But mankind faces a crisis and to overcome it we may be forced to exploit our knowledge of biochemistry in order to imitate living processes. One aspect of the approaching crisis can be seen by looking at the energy problem. Not only does modern man require sources of power to drive vehicles and machines, he also requires energy for the internal bodily activities, and this can be provided only by food.

All our food energy is derived ultimately from the sun. It has been estimated that the sun supplies 5×10^{24} J annually to our planet, and three-quarters of this is absorbed by the atmosphere or is reflected back into space. Of the remainder, perhaps one-tenth of 1 per cent is 'fixed' by photosynthesis.

The demand for energy

The world population is currently estimated as 2.7×10^9 and each individual requires (ideally) about 10^7 J of energy per day. Man's energy requirement can account for as much as 1 per cent of the energy used in photosynthesis. His main energy source, as far as his own bodily needs are concerned, is carbohydrate. Though about half the dry weight of a plant is carbohydrate, most of this is in the form of cellulose and we are unable to digest this material. Though some bacteria can produce an enzyme (cellulase) to catalyse this breakdown, animals cannot do so. Herbivores exploit this bacterial action and digest cellulose indirectly. Noting that our main energy source is carbohydrate other than cellulose we can calculate that nearly one-tenth of the planet's photosynthesis appears to be directed to the maintenance of a single species—man.

The energy crisis

If the present trend continues, the world population will double every 30 years. The present population just manages on existing resources and we can predict an inevitable energy crisis some time towards the middle of the twenty-first century. This will occur even if agriculture and horticulture become very much more efficient than at present. It has been estimated that even if all the conceivably available land was used in the most efficient manner possible, the resulting food energy could be increased only about ten times and this would not be enough to avert disaster.

Energy is not the only commodity at a premium; the quality of man's living is inextricably linked with his exploitation of mineral resources and fossil fuels. Some of these are being used up at a rate which will lead to their complete disappearance in the next 50 to 100 years.

This decline in resources indicates a lack of balance between supply and demand. Man is not managing his affairs well and must reform the techniques he uses to obtain food and other necessary materials if he is to survive. Living organisms (including man) have acquired sophisticated techniques with which to transform the sun's energy and the raw materials of the environment. This sophistication has come slowly, through the action of millions of years of organic evolution.

A question facing us in the near future is whether to go on exploiting the chemistry of living things or to try and imitate the useful bio-chemical processes of nature. By this means man would be able to transfer the sun's energy directly and efficiently into food.

Today, a food factory is concerned with processing natural products—substances made by living organisms. It would be of the greatest value to mankind if methods of making food without recourse to organisms could be devised. The food factory of the future may be able to make food from simple chemical raw materials. Bearing in mind what has been said here about the importance of enzymes, it would seem more than likely that a scheme of synthetic food production would depend upon enzyme action. The use of single enzymes *in vitro* is becoming common but the use of multi-enzyme systems outside the living cell is rare. The prospect of a complex of such systems imitating photosynthesis is a daunting one.

A further step towards complete mastery of the energy and raw materials of this planet would be the synthesis of enzymes themselves. This is not as fantastic an idea as it may sound. Already a practical way of uniting amino acids in specific sequences has been devised. The principle of the Merrifield method of peptide synthesis (named after the inventor) is to treat an amino acid (1) so that the COOH or NH_2 group can be attached to a solid support such as a synthetic polymer (see Chapter 4). After washing away unattached amino acid 1, the system is treated so that the free COOH or NH_2 can be made to combine with another amino acid (2) which is then added. A peptide bond is formed and excess amino acid 2 is washed away. In theory, more amino acids of any kind can be attached, in a specified order, so that long chains can be built up. In practice it is difficult to remove all traces of excess, unattached amino acid at each stage and so obtain absolutely pure samples of the desired product but, doubtless, such difficulties will be overcome in time and we can look forward to the day when proteins and enzymes can be made to order. Such a development would make the idea of *in vitro* photosynthesis less fantastic than at present and as progress in all aspects of science and technology accelerates, so fantasy becomes commonplace.

The hard facts of life are an exploding human population in an environment of limited resources. Man must master all forms of chemical change if he possibly can and the most important agents affecting chemical change, so far discovered, are enzymes. Thus, our very survival as a species on this planet may depend on our knowledge of enzymes.

8.

Techniques of Investigation

Enzymology, like every other branch of science, is partly a body of knowledge, partly a collection of ideas, and also an armoury of techniques for discovering more facts and testing new ideas. The three aspects interact with one another. New facts suggest new ideas and testing them requires experimental techniques. It is just not enough to make statements about enzymes and their supposed activities; statements must be supported by evidence. Evidence is seldom a simple account of observed phenomena but more often involves interpretation of complex experimental results. To interpret properly, one must understand the techniques employed.

A complete study of enzymes involves a progressive unravelling of intact organisms until the chemical activity within them is understood. This means observing the activity of whole organisms, dissecting and sectioning them and further disrupting their cells and organelles so that each active component of the living system is eventually isolated and analysed. (This progressive process is not necessarily carried out by one set of workers in a laboratory but by many different people at different times. The results of these multifarious activities are ultimately put together into a coherent scheme.)

Whole organisms

Some organisms secrete certain enzymes into their immediate environment. Bacteria and yeasts, for example, suspended in a watery medium can act as an enzyme 'solution' and limited studies of the enzyme or enzymes can be carried out without damaging the organisms at all. The majority of enzymes are not secreted freely and, in one way or another, they must be extracted before they can be fully investigated.

Organs

Whole organs, such as kidneys, digestive glands, and livers, can be excised in order to study the behaviour of a single enzyme or complex. Enzymes are not distributed evenly throughout an animal or plant but often located in high concentration in certain organs. By making extracts from isolated organs, the process is simplified and the yield increased.

Cells

Often enzymes are located in particular parts of a cell, associated with organelles such as mitochondria or chloroplasts. Relatively simple histochemical techniques can sometimes reveal them. For example, alkaline phosphatase can be shown to be distributed unevenly in kidney cells. Sections of kidney can be passed through a series of chemical processes so that ions of a cobalt salt become attached to the sites where the enzyme is active and which remain fixed in the sectioned cells. When these are exposed to ammonium sulphide solution a black precipitate of cobalt sulphide forms, indicating very clearly the distribution of phosphatase when viewed under a light microscope.

Electron microscopy

The same principle applies if high magnification and resolution are required and electron microscopy is employed. The specimen is incubated in an enzyme substrate and a soluble salt of a heavy metal such as lead nitrate. The enzyme causes the release of phosphate from the substrate and this reacts to form insoluble lead phosphate which is opaque to electrons. The sites of enzyme activity in the sectioned material can thus be identified.

Quite apart from this specialized technique, the electron microscope has played a prominent part in advancing our understanding of enzyme action within cells. The ultrastructure revealed in micrographs has suggested a physical basis for the highly organized chemistry of the cell. Electron micrographs of mitochondria, for example, show an orderly arrangement of granules attached to membranes which provide just the systematic arrangement necessary for complex, multi-enzyme systems.

Organelles

The next step in the analysis of enzyme systems is the isolation of organelles such as mitochondria, and this can be done only by first disrupting the cells (Figure 8.1). The simplest method is to pound tissue with a pestle in a mortar, sometimes with the addition of sand or other abrasive material. Alternatively, tissue can be forced, in a press, through minute holes, causing disintegration of the cells. More sophisticated methods include bursting the cells by osmotic shock and submitting them to ultrasonic vibrations. Experience has shown that often a shearing rather than a crushing force causes less damage, and *blenders* (similar in principle to domestic homogenizers) have proved successful.

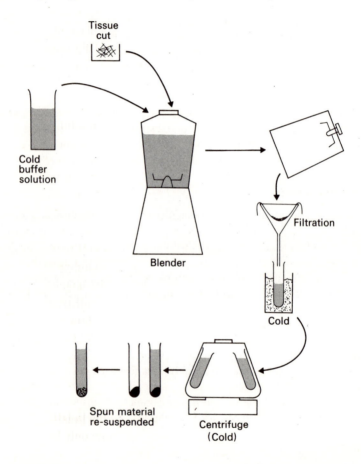

Figure 8.1. Steps in the formation of homogeneous samples of active organelles

Tissue is cut into small pieces in water, with dissolved additives of various kinds, then sharp blades rotate at high speed not only cutting the tissue but causing cavitation which helps to sheer the cells into smaller fragments.

Once the cell membranes are ruptured, all kinds of abnormal enzyme reactions can occur and which must be prevented. For example when an apple is cut, brown tannins form in a short time, owing to the action of enzymes called polyphenol oxidases. This action can be inhibited by the addition of 2×10^{-3} M sodium metabisulphite. Polyvinyl pyrolidone acts as a tannin absorbent. Many such reactions have to be inhibited if serious chemical damage to a preparation is to be avoided, but two general procedures are useful: (1) the cell-breaking process should be done as *rapidly* as possible and (2) it should be carried out at a *low temperature.* Blending and subsequent separations are commonly done in machines incorporating a cooling unit.

Disruption upsets many factors in the cell which may distort normal enzyme action. Any change of pH, for example, will affect proteins and therefore enzymes. For this reason, extractions are commonly done in a *buffer solution,* e.g. phosphate buffer. The situation is further complicated by changes in the osmotic relationships when organelles are thrown out of their cells into a new environment. Substances are added to maintain the proper *osmotic balance* but care must be taken to ensure that such substances (e.g. sucrose) do not, at the same time, act as enzyme substrates so confusing an already complex system. Each known organelle type now has its associated recipes for extraction. For example, the technique for extracting chloroplasts from plant tissue without destroying their ability to fix carbon dioxide is as follows:

Homogenize chilled spinach leaves in medium containing 0.3 M sorbitol, 0.005 M magnesium chloride and 0.01 M sodium pyrophosphate adjusted to pH 6.5. Filter by squeezing through cloth. Centrifuge, 4000 g, rapidly (see below) at 0 °C and re-suspend in a medium containing sorbitol, magnesium chloride, 0.001 M manganese chloride, 0.001 M EDTA, and 0.05 M hydroxyethyl piperazine ethanesulphonic acid with the pH adjusted to 7.6 (procedure of D. A. Walker).

Isolation

A biochemist trying to discover the distribution of enzymes within normal cells must examine and compare the chemical activities of the various organelles and other parts of the cell, which means that all these must be separated from one another, into homogeneous preparations. The most useful tool for this kind of separation is the *centrifuge*. A suspension of cell debris allowed to stand in a test tube would begin to separate out under the influence of normal gravity, the denser material would fall to the bottom but not much separation would be achieved. By spinning the tube in a laboratory centrifuge forces equivalent to 2000–4000 times normal gravity can be achieved. By successive spinning and decanting, fractions containing the different cell constituents can be obtained, provided they differ sufficiently in density. The densities of particles to be separated by this means must differ from one another and from the medium in which they are suspended. Separation can be facilitated if the medium is not uniform but has a *density gradient*. Two liquid media can be used which are immiscible thus forming a discontinuous density gradient across which particles separate. A gentler, continuous gradient can be made using aqueous solutions of such substances as dextrose and sucrose. In theory these should give better separation but, in practice, the discontinuous gradients also give good results.

By thus isolating different parts of cells, the chemical activity of each can be separately investigated and the results pieced together so that ultimately the workings of whole cells can be understood.

Macromolecules

Unfortunately for the investigator, the organelles which can be separated each contain many enzyme systems, and the process of separation and isolation must be continued with particles of much smaller size, in the range between smallest organelles and larger molecules.

Very large molecules like proteins can be separated and isolated in a number of ways depending on differences in *molecular weight* and *electrical charges*. In some respects the techniques used resemble those for larger particles. It is possible to make *sieves* of a kind which sort out protein molecules. They are based on water-insoluble granular materials which are porous and can be organic in origin (seaweed dextran) or synthetic (acrylamide polymer). The pore sizes of the granules are carefully regulated by chemical treatment so that the molecules in solution are excluded from the pores if greater than a certain size; smaller molecules can penetrate the gel. The sample of proteins to be separated is applied to the top of a column of a *gel* (see Figure 8.2) and

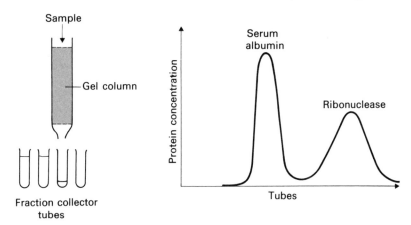

Figure 8.2. Proteins to be separated are applied to the top of a gel column and the fractions are collected in tubes below

eluted down the column, usually with a buffer solution to prevent alteration of the protein molecules. For example, the proprietary molecular sieve made from seaweed dextran called Sephadex is available in a wide range of pore sizes for aqueous and non-aqueous solutions. This will separate serum albumin (molecular weight 63 000) from ribonuclease (13 000). Agarose is another sieve made from seaweed agar; Biogel is made from acrylamide. The serum albumin is excluded from the gel granules but the ribonuclease, being able to penetrate the pores of the gel, is retarded and emerges after the albumin.

Apart from molecular weight, the shape of a molecule is an important factor connected with methods of separation. A thin flexible molecule may be able to penetrate the gel when a spherical, rigid molecule of the same weight cannot do so. Some gels, particularly those made from natural substances, contain polar groupings such as $-COO^-$, $-COOCH_3$, and $-NH_3^+$, which are capable of electrical interaction with proteins, positively or negatively.

Electrical techniques

Polar interactions form the basis of two important techniques for the separation and purification of enzymes and other proteins. The amino acid side chains contain $-COOH$, $-NH_2$ and other groups capable of ionization at certain pH values. Most proteins have a particular pH value at which the number of positively charged groupings (NH_3^+, imidazole) equals the negatively charged ones ($-COO^-$). For ribonuclease, with more basic than acidic groups, this value is around 10; for serum albumin it is around 4. This pH value is the *isoelectric point* of the protein; here an electric field will cause no migration of the protein molecules. At pH 7 ribonuclease and serum albumin will move in opposite directions, ribonuclease with positive charges will migrate to a negative cathode, albumin will move to an anode. This is the basis of a separation technique called *electrophoresis.* Protein solutions in a suitable, conducting buffer have an electric potential applied. The most convenient arrangement has the buffer supported on a strip of paper or other inert material such as cellulose acetate, starch, or synthetic polymer. The outline of the apparatus is shown in Figure 8.3. The protein mixture is applied as a central band using a pipette and a potential of 50 to 200 V is set up across the paper which is saturated with buffer and enclosed so as to retain a saturated atmosphere and prevent drying. After an hour or two the proteins will have separated and their position can be detected by

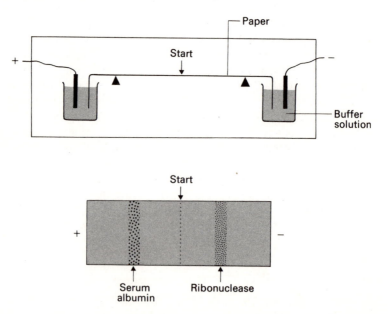

Figure 8.3. Electrophoresis apparatus

using a stain or coloured reagent. Once their position is thus discovered the procedure can be repeated exactly but instead of staining, the portions of paper containing the various separated proteins are cut away and the proteins washed out.

More recently methods have been devised for making *pH gradients* so that better separation of proteins can be affected. Proteins subjected to an electric potential will move unless the pH is that of the isoelectric point. In a pH gradient, therefore, each protein moves until it meets a region where the pH corresponds to its isoelectric point. (There is a similarity in principle between this and centrifuging in a density gradient, described above.)

Another technique used to separate proteins, and therefore of importance in the study of enzymes, also depends on the charged nature of protein side groups. An *ion-exchange column* is a vertical tube containing granules of synthetic resin, or cellulose, treated chemically to give them $-SO_3H$ or $-NH_3^+$ or $-COOH$ groupings. Proteins show an affinity for these materials depending on their charge, the ionic strength of the solution and the pH. The protein is allowed to bind on to the ion-exchange column and is then displaced by an increasing salt concentration or pH or both. The technique can be used to remove single proteins from a mixture and has proved most useful, particularly when treated cellulose is used as resins tend to degrade proteins.

The techniques described, gel filtration, electrophoresis and ion-exchange may be combined together as in ion-exchange Sephadex materials or in polyacrylamide gel electrophoresis. This latter technique is extremely discriminating and is frequently used to check the homogeneity of supposedly purified protein.

The concept of purity is based on the idea of uniform molecular size and structure but it is complicated by compounds which vary greatly in molecular size and electrical charge. A preparation of glycogen, for example, may contain molecules varying in molecular weight from 10 000 to 100 000, and gel filtration is unsuitable for separating this material.

Molecular structure

So far, we have been concerned with techniques which enable the biochemist to find out where enzymes are active within living cells and what reactions they influence. To try and understand how they catalyse their respective substrates some idea or model of the enzyme molecule itself must be established; the structure of chymotrypsin has been mentioned. To make a satisfactory model many lengthy procedures are necessary.

Once a macromolecule has been isolated in a pure state it is necessary to establish its size and shape and initially the molecular weight is established. The three-dimensional structure is deduced from X-ray crytallographic measurements and, coupled with these physical methods, a complete determination of the amino acid sequence is carried out. Further, chemical studies of the active site give additional information on the way an enzyme works. The chemical and physical (X-ray) studies are a useful cross-check on each other.

Molecular weight

Some idea of molecular weight can be gained from gel filtration and reference to pore sizes which exclude the macromolecule and those which do not. Precise determinations of molecular weight require the use of either an *osmometer,* an *analytical centrifuge,* or a means of measuring *light scatter.* All three techniques are theoretically and technically complex.

The *osmotic pressure* of a solution is the pressure required to prevent the entry of pure solvent through an ideal or perfect semipermeable membrane. Alternatively it can be regarded as the 'escaping tendency' of the solute. The larger the molecular weight of the solute the smaller is the osmotic pressure (for a given concentration). It is usually measured as the hydrostatic pressure developed across a solute-impermeable membrane and is effective with solutes of moderate molecular weight but becomes less useful with very large molecules. This is true of the other so called *colligative phenomena* which are related to osmotic pressure, the lowering of solvent vapour pressure, boiling point elevation, and freezing point depression.

Organelles and debris from cell disintegration can be separated by spinning suspensions in a centrifuge. The same principle can be applied to the *sedimentation* of macromolecules to give information about molecular weight. The principle is simple enough but there are great mechanical difficulties in the methods of obtaining the very high gravitational forces necessary. Debris will separate from chloroplasts under a force 1000–2000 times that of gravity (though more rapid spinning at 4000 g is recommended). To cause the sedimentation of a protein in solution requires gravitational forces around 100 000 times

gravity. The machines that can do this are called *ultracentrifuges* and
they achieve speeds of rotation up to 75 000 revolutions per minute. The
rate of sedimentation can be measured (see Figure 8.4) and this can be
coupled with separate measurements (also made in an ultracentrifuge) of
the rate of free diffusion of the protein molecules in solution to give
information about the molecular weight, shape, and size. Alternatively,

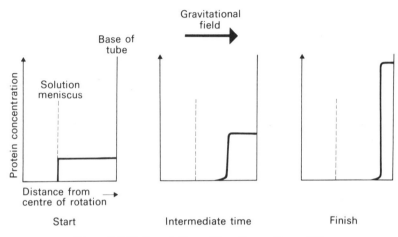

Figure 8.4. Sedimentation achieved in an ultracentrifuge

at a slower speed, the ultracentrifuge can be used to bring about a stable
re-distribution of solute in the solution—the *sedimentation equilibrium*
(see Figure 8.5). This technique, with modern ultracentrifuges, can be
used with substances of relatively low molecular weight (600-2000) such
as sucrose.

If a beam of light is directed into a vessel containing a suspension of
particles such as powdered chalk, the light is thrown out of the vessel in
all directions. If the beam passes into pure water it continues without
any scattering even though the water consists of particles or molecules. It
is the relation between the size of particle and the wavelength of light
which determines the amount of light reflected and scattered. The
theory of *light scatter* is well established and complex. There are
practical difficulties in using it as a technique for investigating proteins
and other macromolecules, chiefly in connection with the very small
amounts of light to be measured and the very large errors that result

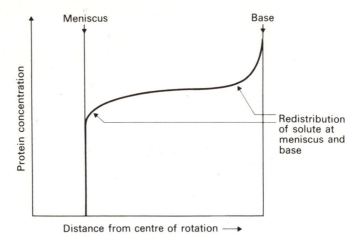

Figure 8.5. Stable redistribution of solute in the solution; the sedimentation equilibrium

from minute impurities and imperfections of the optical systems. The problems have been reduced by the use of *laser beams*, intense beams of collimated, monochromatic light ideally suited to light scatter studies of substances like ribonuclease (molecular weight 13 000).

The usefulness of the various techniques so far described is summarized in Table 8.1.

Table 8.1

Method	Range of measurable molecular weight	
	Minimum	Maximum
Colligative phenomena, elevation of boiling point, depression of freezing point etc.	None	1 000
Osmotic pressure (depending on membrane permeability)	1 000	50 000
Ultracentrifuge—sedimentation equilibrium	500	None
Light scatter	10 000	1 000 000 000*

* Limited by the need for purity and freedom from dust etc.

Molecular shape

Some information about the shape of molecules can be gained from the methods of molecular weight determination described. For example, a sedimentation rate strongly dependent on concentration is indicative of asymmetric molecules which interfere with one another's sedimentation. Osmotic pressure is, in any case, dependent on concentration but if it is observed to be excessively concentration-dependent this indicates asymmetry of molecular shape. The angular distribution of scattered light provides information on structural irregularities of solute molecules.

Various additional techniques are of more limited application. Occasionally *electron microscopy* can provide information about the shape of very large molecules greater than 10^6 molecular weight (p. 15). Measurements of *viscosity* are quite easy to carry out but difficult to interpret. Optical studies of *birefringence* give clues about the orderly or disorderly way in which molecules arrange themselves. All these help to provide a general indication of molecular shape such as 'rod-like', 'spherical', or 'ellipsoidal'. Though this may seem a small reward for expensive and painstaking researches, even these general notions are most valuable when taken in conjunction with other observations of protein and enzyme chemistry.

Atomic arrangement

The structure of an enzyme can be said to be fully known only when the position of every atom and group in the molecule is known. A technique which has been widely used with great success to investigate the fine structure of crystals is *X-ray diffraction.* A crystal exists when a homogeneous collection of molecules is regularly arranged. Though the technique has been known to biochemists for a long time it has not been used because it has been impossible to produce pure samples of biological material which would form crystals. With the improvement in techniques of extraction and purification has come increased use of X-ray diffraction.

Any monochromatic, electromagnetic radiation falling on a regular array of objects (as indicated in Figure 8.6) is reflected from each object. To an observer stationed at O, the lower ray (reflected from B) will travel further than that reflected from A by an amount readily calculated as:

$$2d \sin \theta$$

where d is the distance AB and θ is the angle of incidence of the

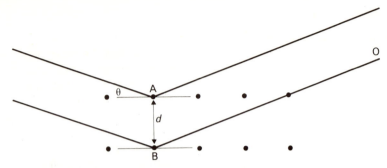

Figure 8.6. Monochromatic radiation reflected from a regular array of objects

radiation. If $2d \sin \theta$ is an exact multiple (n) of the wavelength of the radiation (λ) then the intensity noted by the observer at O will be double that of the single ray. This can be expressed by

$$n = 2d \sin \theta$$

If λ is very much larger than d, no value of θ exists to provide a reflected ray at all; if λ is very much smaller than d then θ is so small that it cannot be separated from the incident beam of radiation. If, and only if, d and λ are of comparable size can realistic values of θ be obtained thus giving the observer a series of maxima and minima. When $2d \sin \theta$ is an odd number of half wavelengths, a pattern of alternate reinforcements and cancellations produces what is known as a diffraction pattern. If there is a large, regular, three-dimensional array of objects, the pattern obtained by an observer (at O) is a two-dimensional arrangement of spots of radiation. From the measurement of these spots, the dimensions of the array can be calculated.

For atomic arrays the value of d is in the range 0.1–0.3 nm so that diffraction patterns are possible only with X-rays, which have a very small wavelength. Light radiation in the visible range, would be far easier to produce and to observe but its wavelength range makes it quite useless for this work. To observe the diffraction of X-rays, photographic film is used; the pattern of spots enables the dimensions of the interatomic distances to be calculated.

Enzymes, and other protein macromolecules, are so complex in their structure, with a thousand or more atoms per molecule, that the X-ray diffraction pattern presents a severe challenge to the investigator. The first macromolecules to be successfully analysed in this way included myoglobin and haemoglobin. These have iron atoms in each molecule and they produce characteristic powerful X-ray reflections. This led to the technique of introducing *heavy metal atoms* into compounds, which normally lacked them, to act as markers and so aid the interpretation of diffraction patterns. Ideally a range of such markers is needed and the technique of inserting them attached to $-SH$ or histidine groups is known as *isomorphous replacement.* The translation of two-dimensional spot patterns into a three-dimensional model of a molecule is not a matter of straightforward calculation. Essentially it consists in setting up hypothetical models, deducing the diffraction that would be caused and comparing this with the actual patterns obtained. This kind of work is well suited to *computers* and they have played a significant part in recent X-ray investigations of organic macromolecules.

The difficulties of purification, crystallization, and heavy metal replacements have restricted the number of compounds investigated to about 20 proteins, including enzymes such as chymotrypsin and ribonuclease. The most famous subject of analysis was DNA, which led Watson and Crick to put forward the double helix structure; more recently, carbohydrates have been studied by diffraction techniques and there is little doubt that it will long remain a most important method of investigation.

Energetics

So far, we have been concerned exclusively with the substances involved in enzyme-controlled reactions and, in particular, with the nature of enzymes themselves. But, important advances in enzymology have come from studies of the dynamics of reaction, the equilibria formed, and the energy gains and losses associated with chemical change. These studies require techniques of quite a different kind from those described above.

Enthalpy

The enthalpy change in a reaction is identical with the heat of reaction (measured at constant pressure); this can usually be determined directly using *calorimetric techniques.* Thermal energy causes a rise in temperature which, multiplied by the mass and specific heat of the material undergoing the temperature increase, provides a measure of the thermal energy. Accurate measurements of the heat produced by a reaction can be very difficult. It is difficult to maintain the reaction and measure the temperature without cooling the system by an unknown amount. Calorimetry has been greatly improved in recent years by the use of electronic devices. Instead of trying to measure the heat change of a single system, two identical systems are set up, one containing reacting material, the other an electrical heating system. Electronic devices ensure that the heat change in the first is exactly balanced by change in the second and the heat change is calculated very accurately from the amount of electricity consumed.

Bond Energy

Bond energies and enthalpies are simply related. Accurate determinations can be made not only from enthalpy measurements but also spectroscopically. Knowledge of bond energy is often critical in the understanding of the mechanism of a chemical change. When two atoms combine to form a chemical bond, the relationship between the bond energy and their distance apart follows the pattern shown in Figure 8.7(a). The diagram indicates that when the two atoms are very far apart (C) no interaction between them is possible. When they come too close together (A) internuclear repulsion makes the energy required unfavourable; a stable bond is associated with the interatomic distance B, or rather a series of states DD', EE', see Figure 8.7(b). Rather than a stable state (B) there is vibration over these interatomic distances. The quantum theory permits only a limited number of such states, interchange between which

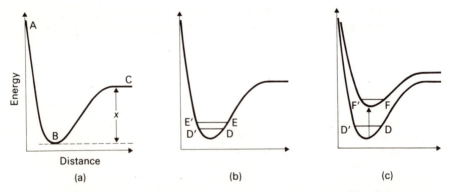

Figure 8.7. Relationship between energy and distance apart of bonded atoms

corresponds to the absorption of specific wavelengths of light or other electromagnetic radiation. The resulting spectra (Figure 8.8) are called vibrational spectra and are associated with the *infra red* region. By extrapolation of such spectra (since infra red radiation causes no molecular dissociation) obtained from an infra red spectrometer, the bond dissociation energy (X in Figure 8.7) can be determined, for the vibrational energy levels tend to converge. Such extrapolated values are rather inaccurate (usually about 10 per cent too high) and, as an alternative to infra red spectra, visible and *ultra violet* radiation can be used. This causes *excitation* from a particular vibrational energy level of one electronic state to a vibrational state in a higher (excited) electronic state (FF'), see Figure 8.7(c). The fine structure of the electronic spectrum does enable much more accurate extrapolation to obtain X, the bond dissociation energy.

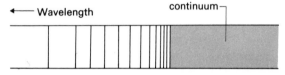

Figure 8.8. Vibrational spectra

Free energy and entropy

The relationship between enthalpy (ΔF) involves entropy (ΔS) thus:

$$\Delta F = \Delta H - T\Delta S$$

(Free energy is commonly symbolized by G and is termed Gibbs' free energy because Willard Gibbs first proposed these relationships.)

   Free energy is related to two measurable quantities, the *equilibrium constant* and the *EMF* of an electrochemical cell. The determination of the equilibrium constant of a chemical reaction is basically one of analytical chemistry, and many methods, microchemical, spectroscopic, thermal, pressure, and volume, have evolved for the evaluation of equilibrium constants. Another technique is illustrated by the study of hexokinase which catalyses:

$$ATP + glucose \rightleftharpoons glucose\text{-}6\text{-}phosphate + ADP$$

Here the equilibrium constant is very large (about 10^6) so at equilibrium very little glucose and ATP are present. By using ADP and glucose-6-

phosphate containing *radioactive carbon* (^{14}C) instead of normal carbon (^{12}C), it can be shown that a little radioactive glucose exists. Methods of measuring the radiation are extremely sensitive so, by isolating glucose from the reaction mixture, the amount present can be accurately determined. The concentration of ATP must be the same as that of the glucose so, knowing the initial ADP and glucose-6-phosphate concentrations the equilibrium constant can be calculated.

Measurements of EMF provide some of the most accurate thermodynamic data. The chief problem in practice is that associated with electrode polarization which distorts the EMF values. Derivation of equilibrium constants from the direct measurements or from EMF values can be carried out over a range of temperatures. The variation of equilibrium constant with temperature provides yet another method for enthalpy changes using the formula

$$\frac{d \ln K}{dT} = \frac{\Delta H}{RT^2}$$

Entropy change (ΔS) can be calculated from measurements of ΔF and ΔH. This probably provides the most reliable basis for determination; absolute (independent) methods for finding ΔS are much more difficult to carry out. For a particular compound the entropy value at any particular temperature T (S_T) is given by:

$$S_T - S_0 = \int_{0°K}^{T} \frac{C_p \, dT}{T}$$

where C_p is the specific heat. Specific heat varies with temperature but the two are not simply related (see Figure 8.9); S_T is the area under the curve, the assumption being that, at $0°K$, $S = 0$ (i.e. the integration constant is zero). This assumption takes into account the *third law of thermodynamics* that at absolute zero the entropy of a perfectly crystalline, pure substance is zero. Entropies obtained from the integration of $C_p \, dT/T$ are known as 'third law entropies'; their prime disadvantage lies in the need to measure C_p at temperatures close to absolute zero, and in this region extrapolation is inaccurate. In addition, any change of phase (state) such as liquefaction or vaporization further complicates the situation, as do structural modifications in the crystal (the validity of the third law for all substances must be doubtful). For a few substances, mainly gases, absolute entropies can be calculated or extrapolated from spectroscopic data. Entropy is a measure of molecular disorder which corresponds to the distribution of a collection of

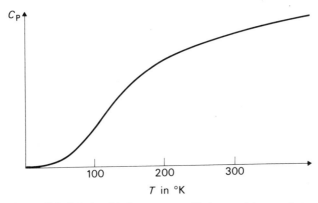

Figure 8.9. Relationship between specific heat and temperature

molecules amongst various energy states—the partition function. In favourable cases this distribution can be calculated if the energy states which involve vibrational (infra red) and rotational (microwave) spectral measurements are known so the absolute entropy can be computed. At present, such data are confined to small molecules.

Scientists investigating enzymes and the reactions they catalyse have a very large, growing armoury of techniques. In addition to all the well established methods of classical chemistry, spectroscopy, and X-ray crystallography, other techniques of physics and biophysics have been added. No one man can, of course, be well versed in all of them. The boundaries within science, dividing it into areas like biology, chemistry, and physics, mean little when discovery is possible only by the interaction of many diverse interests.

Suggestions for Further Reading

Morris, J. G. *A Biologist's Physical Chemistry* (Arnold, 1968)

Bernhard, S. *The Structure and Function of Enzymes* (Benjamin, 1968)

Giese, A. C. *Cell Physiology* (Saunders, 1968)

West, E. S. and Todd, W. R. *Textbook of Biochemistry* (Macmillan, 1966)

Allen, J. M. (ed) *Molecular Organization and Biological Function* (Harper & Row, 1967)

Bartley, W., Birt, L. M. and Banks, P. *The Biochemistry of the Tissues* (Wiley, 1968)

Dixon, M. and Webb, E. C. *Enzymes* (Academic Press, 1964)

Index

ACETATE ION, 23
Acetic acid, 22, 23
Acetyl, CoA, 73
   glucosamine, 64
   phosphate, 41
Acetylcholine, 60
Activation energy, 29, 30, 35, 52
Active sites, 53
Adenine, 87
Adenosine, 43
   diphosphate (ADP), 37
   monophosphate (AMP), 10, 65, 72, 89
   triphosphate (ATP), 37, 63, 72, 83, 109
Agarose, 99
Alanine, 5, 89
Alkaline phosphatase, 95
Allosteric transitions, 70, 72
Allothreonine, 69
Amidine, 58
Amines, 61
Amino acids, 5, 10, 25, 73, 87
Ammonia, 34, 35
Amylase, 12
Amylopectin, 12, 13
Amylose, 13
Anaemia, 75
Arginine, 89
Argon, 18
Aromatic compounds, 21
Arsenic, 59
Ascaris lumbricoides, 91
Asparagine, 89
Aspartic acid, 89

Bacillus subtilis, 45
Base-stacking forces, 26
Benzene, 21, 24, 25
Bilayer, 3
Biogel, 99
Biosynthesis, 38
Birefringence, 105
Blender, 95
Bond energy, 19, 31, 34, 43, 108
Bonds, 17
Buffer solution, 95, 100
Burk, 50, *see* Lineweaver

CADMIUM, 59
Calorimetry, 108
Carbohydrates, 11, 36, 73, 92
Carbon, 18
Carboxypeptidase, 70
Catalase, 29, 46, 49
Cell, membrane, 4
   wall, 3, 4
Cellulase, 92
Cellulose, 3, 12, 13, 92
   acetate, 100
Centrifuge, 95, 98, 102
Centrosome, 4
Chemiosmotic hypothesis, 83
Chlorine, 18
Chlorophyll, 7
Chloroplast, 7, 84, 95, 97
Cholinesterase, 60
Chymotrypsin, 49, 51, 87
Clostridium welchii, 42
Colligative phenomena, 102, 104
Complex kinetics, 68

Copper sulphate, 77, 78
Covalent bonds, 18
Cramp, 74
Cristae, 6
Cysteine, 50
Cytidine monophosphate, 89
Cytochrome, 78, 79, 81, 82
Cytoplasm, 4, 7, 87
Cytosine, 87, 88

DANIELL CELL, 77, 78, 82
Degrees of freedom, 34
Deoxyglucose, 64
Deoxyribose nucleic acid (DNA), 4, 25, 26, 74, 75, 87, 88, 107
Dielectric constant, 20
Diisopropylfluorophosphate (DFP), 59, 60
Double bond, 21

ELECTRODE POTENTIAL, 78
Electromotive force (EMF), 109, 110
Electron microscopy, 95, 105
Electronic spectrum, 109
Electrons, 18
Electrophoresis, 100, 101
Electrovalent bonds, 20
Endoplasmic reticulum, 4, 7, 8, 86
Endothermic reactions, 34
Energy, rotational, 33, 111
    translational, 33
    vibrational, 33, 109, 111
Enthalpy, 31-36, 41, 107, 108, 110
Entropy, 31-36, 43, 109, 110
Enzyme-substrate complex, 46, 50, 61
Enzyme systems, 37
Erythrocytes, 75
Equilibrium constant, 41, 109
Escherichia coli, 1, 3, 4, 5
Ethane, 20
Ethanol, 20, 37
Eukaryotic organisms, 4
Evolution of proteins, 90
External work, 31

FATS, 13
Fatty acid synthetase, 15

Fatty acids, 14, 73
Ferricyanide, 78, 79, 81
Ferrocyanide, 78, 79, 81
Fluorine, 24
Free energy, 30, 32-36, 41, 109
Fructose, 48

GALACTOSE, 12
Gel-filtration, 101
Glucose, 12, 37-42, 48, 49, 63, 76, 82, 109
    oxidase, 45
Glucose-1-phosphate, 39, 41
Glucose-6-phosphate, 39, 41, 42, 49, 72, 109
Glutamic acid, 89
Glutaminase, 42
Glutamine, 42
Glutaminyl transferase, 42
Glycerol, 13, 73
Glycerol-1-phosphate, 41, 42
Glycine, 5, 88, 89
Glycogen, 38, 39, 71
    synthetase, 71, 72
Glycolysis, 82, 83
Golgi body, 4
Guanine, 87
Guanosine monophosphate, 89

HAEM, 67, 68, 69
Haemoglobin, 67-69, 70, 75, 88, 107
Helium, 18
Hexokinase, 38-40, 49, 63, 64
High energy bonds, 41
Histidine, 50, 52, 53, 55, 89
Hydrogen, atoms, 18
    bonds, 22, 26
    electrode, 79
    peroxide, 28-33, 41, 46
    sulphide, 24
Hydrophilic compounds, 25, 88
Hydrophobic bonds, 51, 52
    compounds, 25, 88

INBORN ERRORS (of metabolism), 90
Induced fit, 66
Infra-red, 109

Inhibitors, 59, 61, 62, 64, 70
Internal energy, 31, 36
Invertase, 48, 65, 70
Iodine, 65
Ion-exchange, 101
Ion transport, 39
Isoelectric point, 100
Isoleucine, 69, 70
Isomorphous replacement, 107
Isozymes, 74

KETOBUTYRATE, 69
Koshland, 66
Krebs' cycle, 73, 82, 83

LACTATE, 37, 73, 74
Lactose, 12
Lead, 59
Leclanché cell, 82
Leucine, 69, 88, 89
Ligands, 70
Light scatter, 102, 103, 104
Lineweaver-Burk plots, 50, 61, 62
Linoleic acid, 13, 14
Lipids, 13
Lysosome, 4, 9
Lysine, 89

MACROMOLECULE, 15
Maltose, 11
Membrane, 1, 2, 6, 27, 37, 83-85
Menten, 48, *see* Michaelis
Mercury, 59, 69
Methane, 20
Methanol, 20
Methyl, chloride, 18, 19
    glucose, 64
Michaelis-Menten kinetics, 48
Michaelis constant, 48, 49, 60
Microtubule, 7-9
Mitochondrion, 4, 6, 7, 8, 82, 85, 86, 95
Models, 63
Monoamine oxidase, 61
Multi-enzyme complex, 15, 36, 37, 95
Muscle contraction, 39
Mutation, 88
Myoglobin, 67, 70, 107

NAPHTHALENE, 61
Neon, 18
Nerve gas, 60
Nicotinamide, 65, 82
Nitrogen, 24
Nitrogenase, 35
Norleucine, 69
Nuclei, of atoms, 18
    of cells, 4, 7, 87
Nucleic acids, 10
Nucleolus, 4
Nucleotidase, 65
Nucleotides, 10

OLEIC ACID, 13
Oligomers, 70
Orbitals, 18, 19, 20
Ornithine, 5
Osmometer, 102
Osmotic pressure, 102, 104, 105
Oxidation, 77
Oxygen, 24

PAPAIN, 45
Peroxidase, 45
Pesticides, 60
Phenylalanine, 22, 28, 29, 51, 61, 89
Phenylketonuria, 90
pH, gradients, 101
    optimum, 55
Phosphoenol pyruvate carboxykinase,
    91
Phosphorylase, 71, 72
Phosphorylation, 63
Photosynthesis, 7, 36, 91, 93
Plasmalemma, 2, 7
Platinum electrode, 79
Polar reactions, 44
Polyacrylamide, 101
Polyphenol oxidase, 97
Polyvinyl pyrrolidone, 97
Prokaryotes, 4
Protein synthesis, 5, 8, 92
Proteins, 5, 10, 45, 87
Protoplasm, 1, 2, 4, 7
Purine, 22, 43, 44
Pyrimidine, 22

Pyruvate, 73, 74, 82

RAT, 1
Reaction velocity, 48
Redox, indicator, 80
    potential, 79, 80, 81
Reduction, 77
Resonance, 22
Respiration, 76
Respiratory chain, 82
Ribonuclease, 99, 100, 104
Ribose, 43
    nucleic acid (RNA), 87, 88
    phosphate, 65
Ribosomes, 5, 8, 70, 88
RNA-polymerase, 88

Salix fragilis, 9
Secretion granules, 9
Sedimentation, 102, 103, 105
Sephadex, 99, 101
Serine, 50-52, 60
Serum albumen, 99, 100
Shape, molecular, 105
Sieve, molecular, 98, 99
Sigmoid kinetics, 69
Sodium, 20
    acetate, 21
    chloride, 20
Starch, 12, 100
Stearic acid, 13
Substrate co-operativity, 68
Subtilisin, 45, 58
Sucrose, 36, 48, 65
Sugar, 36

TEMPERATURE, optimum, 58
Template, 63
Thermodynamics, 110
Thiol groups, 59
Threonine, 69
    deaminase (TDA), 69
Thymidine monophosphate, 89
Thymine, 88
Transferase, 71
Tricarboxylic acid cycle, 73, 82, 83
Triglyceride, 13
Triple bond, 21, 22
Tryptophan, 22, 51, 61
Tyrosine, 22, 28, 29, 51, 61, 89

ULTRACENTRIFUGE, 103, 104
Ultraviolet, 109
Uridine, monophosphate, 10, 89
    triphosphate, 39, 40, 71

VACUOLE, 4
Valine, 69, 88, 89
van der Waals' forces, 18, 25
Velocity, maximum, 60
Viscosity, 105

WATER, 24

X-RAY, 102
    diffraction, 105, 106

ZINC, 77, 81
Zymase, 37